Geotechnics and Heritage: Historic Towers

Renato Lancellotta
*Department of Structural, Geotechnical and Building Engineering,
Politecnico di Torino, Turin, Italy*

Alessandro Flora & Carlo Viggiani
*Department of Hydraulic, Geotechnical and Environmental Engineering University
of Naples Federico II, Naples, Italy*

CRC Press is an imprint of the
Taylor & Francis Group, an **informa** business

A BALKEMA BOOK

CRC Press/Balkema is an imprint of the Taylor & Francis Group, an informa business

© 2018 Taylor & Francis Group, London, UK

Typeset by MPS Limited, Chennai, India
Printed and Bound by CPI Group (UK) Ltd, Croydon, CR0 4YY

All rights reserved. No part of this publication or the information contained herein may be reproduced, stored in a retrieval system, or transmitted in any form or by any means, electronic, mechanical, by photocopying, recording or otherwise, without written prior permission from the publishers.

Although all care is taken to ensure integrity and the quality of this publication and the information herein, no responsibility is assumed by the publishers nor the author for any damage to the property or persons as a result of operation or use of this publication and/or the information contained herein.

Library of Congress Cataloging-in-Publication Data

Applied for

Published by: CRC Press/Balkema
 Schipholweg 107C, 2316 XC Leiden, The Netherlands
 e-mail: Pub.NL@taylorandfrancis.com
 www.crcpress.com – www.taylorandfrancis.com

ISBN: 978-1-138-03272-9 (Hbk)
ISBN: 978-1-315-38746-8 (eBook)
ISBN: 978-1-138-74867-5 (Pbk)

Geotechnics and Heritage: Historic Towers – Lancellotta, Flora & Viggiani
© 2018 Taylor & Francis Group, London, ISBN 978-1-138-03272-9

Table of contents

Preface vii

On the importance of towers 1
Tomaso Montanari

A first insight into towers' behaviour: Geotechnical and structural mechanisms,
leaning instability, long-term behaviour 5
A. Flora, R. Lancellotta, D. Sabia & C. Viggiani

Tower of Pisa: Lessons learned by observation and analysis 15
M. Leoni, N. Squeglia & C. Viggiani

The Ghirlandina Tower in Modena: An example of dynamic identification 39
R. Lancellotta & D. Sabia

The Big Ben Clock Tower: Protective compensation grouting operations 59
D.I. Harris, R.J. Mair, J.B. Burland & J.R. Standing

Preservation of historic towers in Venice: The instructive monitoring-driven intervention
on the foundations of the Frari bell tower 73
G. Gottardi, M. Marchi, A. Lionello & C. Rossi

Carmine Bell Tower in Napoli: Prediction of soil–structure interaction under seismic actions 99
F. de Silva, F. Ceroni, S. Sica & F. Silvestri

The Leaning Tower of St. Moritz: A structure on a creeping landslide 123
A.M. Puzrin

The towers of St. Stephen's Cathedral, Vienna 145
H. Brandl

The tower of the Admiralty in Saint Petersburg 171
V. Ulitsky, A. Shashkin, C. Shashkin & M. Lisyuk

Preservation of the main tower of Bayon temple, Angkor, Cambodia 191
Y. Iwasaki & M. Fukuda

Safeguarding the leaning Minaret of Jam (Afghanistan) in a conflict scenario:
State of the art and further needs 229
A. Bruno, C. Margottini, L. Orlando & D. Spizzichino

Author index 261

Geotechnics and Heritage: Historic Towers – Lancellotta, Flora & Viggiani
© 2018 Taylor & Francis Group, London, ISBN 978-1-138-03272-9

Preface

The preservation of monuments and historic sites is one of the most challenging problems facing modern civilization because of the inextricable mix of different factors: cultural, humanistic, social, technical, economical, administrative.

The awareness of the role of Geotechnical Engineering in such a context moved Arrigo Croce and Jean Kerisel to propose the institution of an international Technical Committee on these topics. The ISSMFE, now ISSMGE, did establish in 1981 the TC19 – Preservation of Historic Sites; since 2001 the Committee has been renamed TC301 – Preservation of Monuments and Historic Sites. Initially, the Committee was supported jointly by the French and Italian National Geotechnical Societies; since 1989, by the Italian Society. It has been chaired by Jean Kerisel, Arrigo Croce, Ruggiero Jappelli, Carlo Viggiani and Renato Lancellotta.

The Committee organized two International Symposia on Geotechnical Engineering for the Preservation of Monuments and Historic Sites, both held in Napoli in 1996 and 2012. They were attended by over two hundred people presenting over one hundred papers, now collected in two nice volumes of Proceedings each totalling almost one thousand pages. TC301 has also published a special volume on Geotechnics and Heritage (2013) and promoted the Kerisel Lecture, held for the first time by Giovanni Calabresi at the XVIII ICSMGE in Paris, 2013.

Experience gained over the years proves the vulnerability of cultural heritage to a variety of geotechnical risks. This poses the problem of conceiving and implementing remedial measures preserving that immaterial value represented by the integrity of the single monument or the "diffuse cultural heritage", that is the goal of any conservation intervention. In addition, it must be kept in mind that each monument or historic site is an "unicum" in itself, neither repeatable nor reproducible. It is then evident how difficult, or even to some extent dangerous, providing guidelines or recommendations with the aim of developing some sort of code of practice could be.

What is actually needed is a conceptual framework, based on a shared culture between geotechnical engineers and professionals as art historians, archaeologists, restorers, architects. This need has inspired the idea of relying on well documented case histories as the most effective means to build a consolidated and shared lexicon and to highlight the basic principles by learning through real cases.

Such an idea lay at the heart of the publication by TC301, in 2013, of the special volume "Geotechnics and Heritage"; the present volume, "Geotechnics and Heritage: Historic Towers", moves along the same path. In this case, well documented case histories of towers (i.e. slender structures transmitting high loads to the subsoil) of historic relevance that had or have geotechnical problems have been collected. The outcome presents an interesting picture of the variety of geotechnical problems that an historic tower may have to face in its life, and of the lack of unicity of solutions for the preservation actions to be implemented. Different solutions have been adopted in the reported case histories, reflecting also the existence of different cultural approaches to the topic around the world.

We hope that this undertaking will contribute to encourage geotechnical engineers to devote all possible efforts to the conservation of historic towers. All over the world, they are not only a distinctive feature of old towns and sites; representing a part that stands for the whole, they generate a deep consciousness of belonging to a community. A good reason to care about them and take good care of them.

The Editors

Geotechnics and Heritage: Historic Towers – Lancellotta, Flora & Viggiani
© 2018 Taylor & Francis Group, London, ISBN 978-1-138-03272-9

On the importance of towers

Tomaso Montanari

To defend, not to attack. Anyhow to see further afield. To challenge the sky or to simply observe it. To call to prayer or to sound the alarm. To do anything an individual can do to serve its community. To look inside oneself and lift oneself above and away from the struggles of daily life. A tower can be used for all these things, a creation which sews together East and West and has its roots in the Bible and the Koran (*Wheresoever you may be, death will overtake you even if you are in tower built up strong and high.* Sura IV, v. 78) and indeed in the origins of our common civilization.

In the Book of Genesis the city and the tower are born simultaneously, they are a unique plan where the universal horizontality of men with a common language is indistinguishable from the verticality of the relationship with the heaven. A relationship which is compromised as it's weighed down by humanity. It is men who climb the sky with the tower, forcing God to come down and see it (*Genesis 11:4 Then they said "All the Earth had unique language and unique words". Emigrating from the East men arrived to a plain in the Sinar region and there they settled. They said to each other: "Come, let us make bricks and cook them in fire". They used the brick as a stone and the bitumen as mortar. Then they said: "Come, let us build ourselves a city, and a tower with its top in the heavens, and let us make a name for ourselves; otherwise we shall be scattered abroad upon the face of the whole earth." [11:5] But the Lord came down to see the city and the tower, which mortals were building.*) This outcome is not inverted in western consciousness until that milestone, the dedication of *de Pictura* by Leon Battista Alberti to Filippo Brunelleschi (1435), where he celebrates not a tower but a dome which not only reaches the sky but surpasses it. (*Who could ever be hard or envious enough to fail to praise Pippo the architect on seeing here such a large structure, rising above the skies, ample to cover with its shadow all the Tuscan people, and constructed without the aid of centering or great quantity of wood?*)

The tower, however, is not only a symbol of the man's ambitious search for knowledge, it's also an architectural masterpiece, a famous example of beauty. Bearing this in mind we see it mentioned in other pages of the Bible (*Thy neck is like the tower of David built with turrets, whereon there hang a thousand shields, all the armours of the heroes…Thy neck is as a tower of ivory; thine eyes as the pools in Heshbon, by the gate of Bath-Rabbim; thy nose is like the tower of Lebanon which looketh toward Damascus. I am a wall, and my breasts like the towers thereof; then was I in his eyes as one that found peace* (Song of Songs: 4.4; 7.5; 8.10)).

The "Ivory Tower": a metaphor which is indeed full of future. It is to it that Erwin Panofsky dedicates his inaugural address which was read at the opening of the academic year at Harvard in 1957. Here Panofsky points out the moment in which the image in the Song becomes "*the symbol of a man withdrawing from active life and social responsibilities into a state of intellectual seclusion*". Write Panofsky "*The story begins, it seems, quite late in 1837 when the French poet Charles-Augustin Sainte-Beuve, in his Pensées d'Aout, contrasts Victor Hugo who upholds the banner of his political creed in battle, with the reserved Alfred de Vigny who, though sharing Hugo's convictions "withdraws before noon as though into his ivory tower"*. A theme to which Henry James also dedicates his last unfinished novel. The Ivory Tower, precisely. In his memorable conclusion Panofsky brings back the most sophisticated meaning of the tower, "*the tower of seclusion, the tower of selfish bliss, the tower of meditation – this tower is also a watchtower. Whenever the occupant perceives a danger to life or liberty, he has the opportunity, even the duty, not only to signal …but also to yell, on the slim chance of being heard, to those on the ground. Socrates, Erasmus of Rotterdam, Sébastien Castellion, Galileo, Voltaire, Zola, the seven professors of Gottingen, Albert Einstein – all tower dwellers if there were any – have raised their voices when they felt that there was danger for the liberty. And though these voices were often ignored or even silenced at the time, they continue to ring in the ears of posterity".*

In conclusion, I would like to return to *Song of Songs* in order to find out what the so-called ivory tower really was. It is not only in the seventh but also in the fourth chapter that the neck of the lady beloved is likened to a tower. But in this other passage there is no transference of attributes; here the tower is not referred to as a tower of ivory; it is described as a structure, even more formidable than a mere watchtower: *"Thy neck is like the tower of David, built for an armoury, whereon there hang a thousand bucklers, all shields of mighty men"*. It is for the mighty men to get the shields and use them in battle. The watchman can only sound the alarm. But in order to do at least that much, he should stick to the tower.

We all know, however, that towers can collapse. Indeed, it is precisely this tragic failure, like many others in history, that inspired one of the conversations of the Gospel that, used over the course of many centuries, tries to find a sense in multitude of catastrophes which occur to the architectures of the man. *"Or those eighteen, on whom the tower in Siloam fell, and killed them; do you think that they were worse offenders than all the men who dwell in Jerusalem?"* Jesus Christ asks his disciples (Luke, 13.4).

The indestructible tower, on the contrary, that of solid foundations and stable top, becomes a paradigm of sound morals; in the fifth song of Alighieri's *Purgatory* Virgil orders Dante to *"stand firmly like a tower, whose high top from blast of wind will never oscillate"*.

There is all this and much more in the consciousness of men of the XIX century, and it is the reason why they take care of the towers they inherited.

At 9.52 am on the 14th July 1902, the bell tower of San Marco in Venice collapsed leaving a pile of debris more than 20 m high. In a parliamentary debate the minister of education, Nunzio Nasi, then responsible for *Fine Arts,* under pressure from the deputy Pompeo Molmenti, pronounced these words of wisdom: "the government can only respect the wishes of the Venetians". And on the 19th July the government representative announced that the bell would be rebuilt, "exactly as it was and where it was"; it was unthinkable that Venice could lose its profile. They even printed a stamp with that motto and in 1908 the new bell tower was inaugurated.

It was in the face of the colossal destruction of the Second World War that the motto became relevant again. In April 1945, the art historian Bernard Berenson writes "If we love Florence as a historical entity, which has lasted through the centuries, as a configuration of forms and profiles that has remained unusually intact despite the changes man has imposed upon it, then we have to rebuild it just as we rebuilt the bell tower of San Marco, *"exactly as it was and where it was"*. And it was this spirit, connecting the rebuilding of the city and the reconstruction of democracy, which led to Article 9 of the Italian Constitution: *The Republic…shall protect the landscape and the Nation's historical and artistic heritage.*

For this reason, the explosives which, in 2012, demolished the towers and campaniles of Emilia, already damaged by the earthquake and hastily judged to be beyond repair, was a terrible signal. The same terrible fate beset that of the bell tower of Amatrice which hadn't been shored up properly after the earthquake of 2016 and was consequently destroyed by that of 2017.

It seems that there remains only one type of tower in our hearts, with that which Salvatore Settis called *"the rhetoric of the skyscraper"*: the 250 m high tower that the stylist Pierre Cardin wanted to erect on the side of the Venice lagoon, or perhaps the incredible vertical cemetery including more than 20,000 loculi that the local government of Verona intends to build.

The basic difference between this type of modern tower and the ancient ones is that the latter ones weren't conceived in opposition to the land on which they were built; going upwards, they didn't break but united a landscape. Like the tower that Pliny the younger had built in his villa, *"From thence you go up a sort of turret which has two rooms below, with the same number above, besides a dining-room commanding a very extensive look-out on to the sea, the coast, and the beautiful villas scattered along the shore line. At the other end is a second turret, containing a room that looks out on the rising and setting sun".*

For this reason, while we ask ourselves what is the meaning of the undertaking of the multinational company, building a copy of San Gimignano (and all its towers) in the suburbs of Chongqing, one of the biggest megalopolis of the world, we believe that it is fundamental for our children, before the moon is extinguished, that they see the *real* San Gimignano with its fourteen towers. And it is for this reason that I, personally, the art historians of the world, and indeed the majority of our citizens, are not only admirers of the Authors of this book, but are deeply grateful to them for the contents of the following pages.

Because there is something that we tend to forget, and it's that if we don't study and take care of these towers they may only exist in our metaphors. But without being able to actually see them, live in or touch them, these metaphors would be silenced.

Geotechnics and Heritage: Historic Towers – Lancellotta, Flora & Viggiani
© 2018 Taylor & Francis Group, London, ISBN 978-1-138-03272-9

A first insight into towers' behaviour: Geotechnical and structural mechanisms, leaning instability, long-term behaviour

A. Flora
Department of Hydraulic, Geotechnical and Environmental Engineering, University of Naples Federico II, Naples, Italy

R. Lancellotta & D. Sabia
Department of Structural, Geotechnical and Building Engineering, Politecnico di Torino, Turin, Italy

C. Viggiani
Department of Hydraulic, Geotechnical and Environmental Engineering, University of Naples Federico II, Naples, Italy

1 INTRODUCTION

There is an unbelievable number and variety of towers in the ancient and recent history of humankind: from the mythical tower of Babel, erected by Nebuchadnezzar in the 6th century BC, and the lighthouse of Alexandria, one of the seven wonders of the antiquity, to the Eiffel Tower, the Burj Khalifa and other skyscrapers, the modern towers for communication.

There were around 7,000 prehistoric drystone round towers (*nuraghi*) in Sardinia, and at least 300 of them are still there. In the city of Bhubaneswar, one of many hundreds in India, there are around a thousand tower temples. In the Île-de-France there are more than 1,200 gothic towers.

Italy is rich in medieval towers. The fresco of Saint Theodore protecting Pavia (Figure 1) is an impressive and remarkable picture of a skyline dominated by tall private towers (over 40 built in the Middle Ages). This was also the case in Bologna, San Gimignano, Florence, Lucca, Pisa, Genova, Ascoli Piceno, Pistoia, Bergamo and many other Italian cities during the 10th and 11th centuries, when an impressive development of private towers took place as a tangible sign of the economic and social power acquired by individual citizens and merchants. The skyline of Florence was dominated by some 150 towers; in Bologna, private towers reached around 180 in number. In San Gimignano, about 60 km south of Florence, there is evidence that 72 towers were present in the 13th century, when the city lay along the *Via Francigena*; 14 of them survive (see Figure 2) and give us an idea of the original character of the urban environment, so distinctive as to be referred to as "the civilisation of the towers".

All the ancient cities of Europe are similarly adorned by extraordinary medieval towers. In 1643, Matthäus Merian the Elder published the volume *Svevia* of his *Topographia Germanica*, which included many engravings of such cities as Nördlingen, Ulm, Rottweil, Gdańsk, Nuremberg and Cologne. Figure 3 depicts the city of Ulm as drawn by Merian, together with a recent photograph from the same viewpoint.

There are 350 years between the two images, and in the intervening centuries Ulm has been densely built up; in spite of this, the cathedral still presides over the modern industrial city (Heinle & Leonhardt, 1988).

In Christianity, since the 6th century, many churches and practically all monasteries have had a tower, or a campanile. In Islam, the muezzin calls to prayer from a minaret. In Chapter 12 of this volume, Bruno *et al.* report on the extraordinary minaret of Jam, in Afghanistan, rediscovered in 1944. In the 10th century, Chinese pagodas had already reached a height of 150 m.

Only in the last century have Europeans recognised the splendour and majesty of the ancient temples of India, Nepal, Burma and Cambodia. Many of them are tower-shaped, and fascinate above all for the magnificence of the mythological representations sculpted in their stones. In Chapter 11 of this volume, Iwasaki reports on one of these.

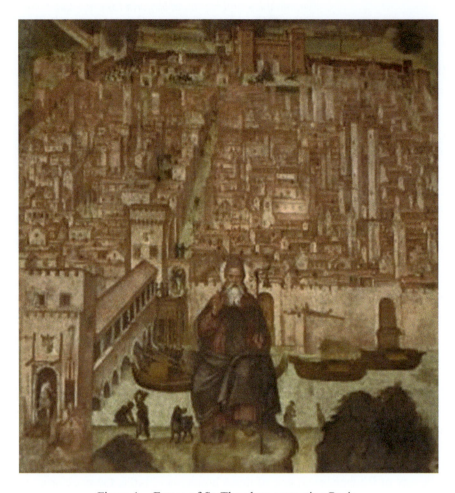

Figure 1. Fresco of St. Theodore protecting Pavia.

Figure 2. San Gimignano cityscape.

Many of these ancient towers are affected by geotechnical and structural problems. A number of representative case histories are collected in the present volume, intended as a contribution to the topic of the preservation of historic sites by Technical Committee 301 of the International Society for Soil Mechanics and Geotechnical Engineering.

Figure 3. The city of Ulm in an engraving by Matthäus Merian the Elder (1643) (top) and in a photograph of 1984 (bottom).

2 LACK OF STRENGTH: GEOTECHNICAL AND STRUCTURAL MECHANISMS

Due to their slenderness, the high stresses acting in their structures and foundations, and the effects of lateral actions such as wind and earthquakes, many ancient towers resting on weak soils have suffered collapses. The most illustrious example might be that of the tower of Babel, which was a ruin in the 4th century BC when Alexander the Great conquered Babylon. Today, the bearing capacity of a foundation may be reliably evaluated; for instance, it is well known that plasticity theorems help us to define the bounds of collapse load, by use of the *static* and *kinematic* approaches. When such an evaluation is carried out for historic towers founded on soft clays, we realise that during the first stages of construction many of them could have been close to an undrained bearing capacity failure, due to *lack of strength* of the soil. Some towers actually survived thanks to delays or interruptions in the building process, which allowed consolidation of the foundation soils with excess pore pressure dissipation and improvement of their strength; the towers could then be successfully finished. Such a process is well documented, for instance, for the leaning Tower of Pisa and the Ghirlandina tower in Modena, both described in detail in Chapters 3 and 4, respectively, of this volume.

It appears, at least in the case of Pisa, that the interruptions in the construction sequence were not intended as a precautionary measure, but were merely fortuitous. Indeed, comprehension of consolidation phenomena was beyond the reach of a medieval mason; it cannot be ruled out, however, that on a merely empirical basis a stop to improve the static situation was routinely practised.

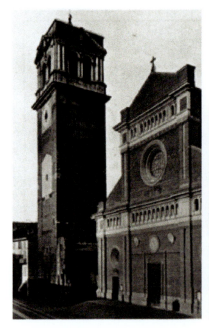

Figure 4. The Civic Tower of Pavia before its collapse (left) and following the collapse of 17 March 1989 (right).

A more subtle and often unforeseen danger was represented by some aspects of a structural nature, linked to damage of the masonry apparatus and stress concentration on masonry contact points. The fatal collapse of the Civic Tower in Pavia, for instance, is a paradigmatic example in this context. On the 17 March 1989, the 900-year-old Civic Tower suddenly collapsed without any forewarning (Figure 4). Investigations showed that environmental effects (i.e. deterioration of masonry due to physicochemical attack) were limited to a few centimetres under the facing (Macchi, 1993). Further investigation of the possible existence of a structural deficiency, based on mechanical tests on masonry blocks and stress analysis, led to the conclusion that the compressive strength of the masonry had been reached on several adjacent elements near the staircase at the base of the tower.

It was therefore concluded that the collapse was triggered by an excess of load in comparison to the very low strength of the masonry, (from 2.6 to 3 MPa), typical of many medieval structures (Binda *et al.*, 1992; Macchi, 1993).

Additional causes of local damage or failure are linked to deterioration of the external facing, as well as of the mortar. Furthermore, we have to consider reduction of the resistant horizontal section along the axis of the tower, due to reduction of the wall thickness or the presence of openings. All these aspects deserve special attention because they can induce local instability phenomena or weakness points that are difficult to forecast because of the absence of any advance evidence.

The recent earthquake in the Emilia Romagna region (May 2012) is a relevant example in this respect, because of the collapse of many campaniles, despite the moderate magnitude ($M = 5.9$) of the seismic event. And, in this respect, it is important to outline that the availability of measurements of acceleration spectra along the axis of a tower greatly enhances our capability to understand and analyse its behaviour and hence preserve the local and global integrity of the structure.

It has been observed that, because of the slenderness of the towers, even in static conditions there are high stresses in the masonry walls and in the foundation soil. In addition, because very often the tower is an integral part of larger buildings, its masonry ties in or interacts with the connected buildings, so that there are horizontal constraints that radically change its behaviour, by reducing its slenderness or introducing concentrations of stresses and additional eccentricities with a very important impact, especially in the presence of seismic actions. For these reasons, there are recurrent damage mechanisms and failures that are dependent on geometrical characteristics, construction methods and the quality of materials, as well as the state of preservation, as unfortunately confirmed by the experience gained during major Italian earthquakes. The images shown in Figure 5, with their accompanying remarks, summarise the most commonly recurring mechanisms.

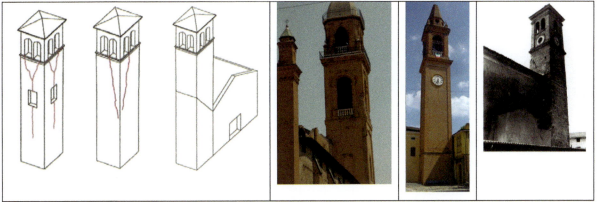

(a) Vertical or inclined cracks as a result of bulging of one or more corners. Shear cracks at the connection between a bell tower and a church, typically due to local hammering or changes in stiffness.

(b) Cracks in the arches and rotation and sliding of the piers.

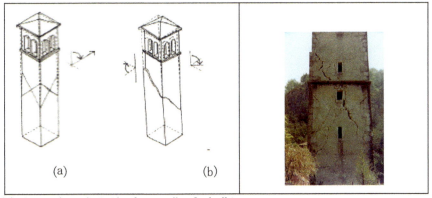

(c) Diagonal cracks in the four walls of a bell tower.

Figure 5. Common failure mechanisms in masonry towers caused by earthquakes.

3 LACK OF STIFFNESS: LEANING AND ASYMPTOTIC INSTABILITY

The fascination of a leaning tower actually serves to hide problems related to its interaction with the foundation soil, which are sometimes not so evident at first glance. At the same time, we can note how the observer, even if not aware of any relevant scientific knowledge, nevertheless has the ability to capture the essence of the problem because, in asking themselves about the reaction of the tower if perturbed by an external action, they are effectively debating its *stability of equilibrium*.

At first glance, a convenient way to explore the leaning instability mechanism is to make reference to the inverted pendulum model (Figure 6). The tower is represented by a rigid column of length h, connected at its base to a rotational spring, representing the soil–foundation system. It is then apparent how the danger of a *leaning instability* increases if there is *lack of stiffness* of the soil, and in this respect the leaning Tower of Pisa represents a powerful example, because the preservation of the tower was recognised as being a problem of leaning instability (Hambly, 1985; Lancellotta, 1993; Como, 1993; Desideri & Viggiani, 1994;

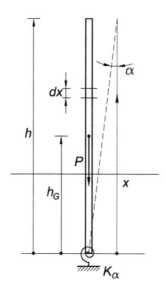

Figure 6. The inverted pendulum model.

Nova & Montrasio, 1995; Federico & Ferlisi, 1999; Burland *et al.*, 2003; Marchi *et al.*, 2011). It is also apparent, therefore, that the crucial step in the analysis is to establish a reliable response for the tower foundation.

In Chapter 3 of this volume, Leoni *et al.* adopt as a first approximation the model of a rigid circular plate over elastic half space, and define a safety factor (*FS*) against instability as:

$$FS = \frac{k_\alpha \alpha}{Wh \sin \alpha} = \frac{ED^3}{6(1-\nu^2)} \frac{1}{Wh}$$

In undrained conditions and in terms of total stress the modulus E_U of a poroelastic body is $E_U = 3E/2(1+\nu)$, and the Poisson ratio $\nu_U = 0.5$. It follows that in passing from undrained to fully drained conditions, the safety factor against instability decreases by as much as 33%. Even this simplistic linear model explains the decrease of stability with elapsing time; of course, if non-linearity and creep are accounted for, the effect may be much more significant.

When dealing with tall structures, the complexity increases due to the interaction between a number of geotechnical and structural phenomena. Initial imperfections and non-uniform foundations very often generate differential settlements which, coupled with low stiffness of the soil–foundation system and delayed deformations, can lead to their long-term critical condition. However, depending on soil stiffness and viscosity properties, we can reach a long-term stable condition just as well as we can approach an unstable one.

4 CAPTURING TOWER BEHAVIOUR TO PRESERVE ITS INTEGRITY

Preserving the integrity of a monument requires a multidisciplinary approach as well as the development of an attitude of not rushing to decisions about stabilising measures until the behaviour of the monument is properly understood (Calabresi & D'Agostino, 1997; Calabresi, 2013).

The case of Pisa Tower, as well as that of Frari bell tower in Venice, discussed in detail in Chapters 3 and 6, provide relevant examples in this respect. They show how long-term monitoring was of benefit in assessing the need for stabilisation measures. The cases of St. Moritz tower (Chapter 8) and of the four towers of St. Stephen's Cathedral in Vienna (Chapter 9) also highlight how important it is to know as much as possible of the whole history of a tower and of the preservation measures implemented over time.

In the case of Pisa Tower, because it is so well known (Burland & Viggiani, 1994), we can read the history of the tilt both in the adjustment made to the masonry levels during its construction and in the shape of the axis

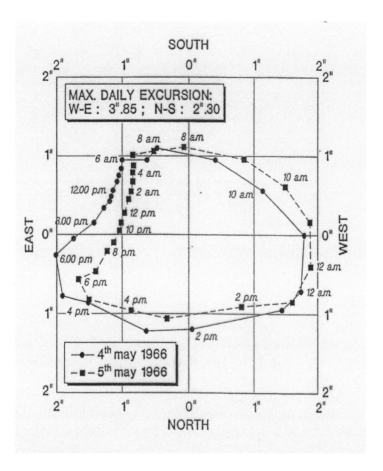

Figure 7. Measurement of the daily variations of the inclination of the Tower of Pisa (Burland & Viggiani, 1994).

of the tower. For the Ghirlandina Tower in Modena (see Chapter 4), there is similar evidence that the tower began to tilt during its construction, because the masons corrected the verticality of the walls accordingly.

The continuing movements of historic towers tend to be seasonal and even daily (see Figure 7); for this reason, it is again very important to have reliable measurements over a long-term period, in order to distinguish between components of differing frequency. Sensors have to be positioned in such a way that any movement should give rise to a consistent set of readings in different locations, thus providing redundancy of information.

An aspect that deserves special attention is the possibility of a difference between the movement of a structure and that of its foundation. This was investigated with particular care for the Pisa Tower (Figure 8), and the high-precision levelling measurements helped to clarify the source of the continuous tilting of the tower. This provided a sound basis for the analysis of leaning instability and suggested that even a small decrease in the tower's inclination would result in a substantial improvement against the phenomenon of instability.

In order to consider all possible critical mechanisms on towers, it is of the utmost importance to monitor environmental conditions such as water table fluctuation, temperature, solar radiation and wind. As an example, it was observed that peaks in water level corresponding to intense rainfall events produced instantaneous southward tilting of the Pisa Tower, with an increment that was not completely recovered. The gradual accumulation of such small, irreversible rotations was believed to be one of the main factors for the steady increase of the inclination experienced by the tower in the past, and this conclusion suggested a drainage system north of the tower to mitigate the effects of the water table fluctuations (Burland *et al.*, 2003).

A very detailed monitoring of environmental conditions was also carried out at Angkor (Cambodia) in an effort to control their effects on the foundation embankment and the masonry structure of the central main tower of Bayon Temple (see Chapter 11).

However, forecasting tower behaviour, as well as its response to seismic actions, requires the establishment of a reliable response by the associated soil–foundation system (Gazetas, 1991), as well as assessment of

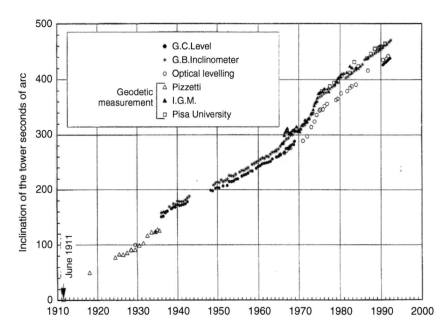

Figure 8. The increase in the inclination of the Tower of Pisa (Burland & Viggiani, 1994). Engineer's levels and optical levelling measure the inclination of the base of the tower; other measurements include deformations of the tower body.

the tower's structural properties. By considering the 3D nature of the problem, the mechanical heterogeneity induced by the stresses applied, and the effectiveness of the soil contact along the vertical sides of the embedded foundation (Pisanò et al., 2014), an effective approach derives from experimental identification analyses, which allow not only estimation of the dynamic parameters of the structure, such as natural frequencies of vibration, mode shapes and damping, but also properly take into account its interaction with the foundation soil, as will be highlighted in Chapters 4 and 7.

A new challenge for historic towers is posed by the continuous transformation of the surrounding built environment, which may affect their static conditions. Modern urban needs, such as new subway lines, have to face the requirement to fully preserve the integrity of monuments. The case study of the Big Ben clock tower in London (Chapter 5), for which compensation grouting was adopted to minimise damage caused by the construction of the Jubilee Line Extension station at Westminster, is a good example in this respect, as it demonstrates the success of this protective measure and shows the value of high-quality monitoring and of numerical modelling. A different use of numerical modelling is shown in Chapter 10 with reference to the tower of the Admiralty in St. Petersburg. Complex 3D numerical modelling was used in this case to identify potentially critical mechanisms, more than to design protective interventions. The construction of a subway line close to the tower was also identified as one of the causes of the observed movements of the tower. Another very interesting example of interaction between a tower to be protected and a new underground construction is that of the towers of St. Stephen's Cathedral in Vienna, described in Chapter 9. A new metro line was excavated close to the cathedral in the 1970s, along with a four-storey underground station. In this case, the design philosophy was to avoid any intervention on or below the cathedral and its towers to preserve the integrity of the monument, thus necessitating extreme care in the design and execution of the immediately adjacent retaining structures. In this instance, the design choices were completely successful in avoiding any effect on the cathedral, thus confirming – along with the examples described in Chapters 5 and 10 – that urban transformation is not necessarily in conflict with the preservation of heritage.

All the relevant examples recounted above demonstrate that a sound interpretation of geotechnical and structural data, as well as long-term monitoring, help in understanding the possible critical mechanisms, eventually leading to the definition of preservation interventions on a rational basis. However, this is not always possible. An enlightening example in this sense is given by the leaning minaret of Jam (Afghanistan), discussed in Chapter 12. This fascinating minaret – one of the tallest in the world – is located in a remote territory, extremely difficult to reach with even the simplest of the machines needed for *in situ* geotechnical testing. Therefore, evidence of a risk of collapse had to be faced without the possibility of drilling boreholes or carrying out the most typical *in situ* geotechnical tests. In this case, such criticalities were, nevertheless, overcome by using less usual investigation techniques.

REFERENCES

Binda, L., Gatti, G., Mangano, G., Poggi, C. & Sacchi Mandriani, G. (1992). The collapse of the Civic Tower of Pavia: A survey of the materials and structure. *Masonry International*, 20(6), 11–20.

Burland, J.B. & Viggiani, C. (1994). Osservazioni sul comportamento della Torre di Pisa. *Rivista Italiana di Geotecnica*, 28(3), 179–200.

Burland, J.B., Jamiolkowski, M. & Viggiani, C. (2003). The stabilization of the leaning Tower of Pisa. *Soils and Foundations*, 43(5), 63–80.

Calabresi, G. (2013). The role of geotechnical engineers in saving monuments and historic sites. Kerisel Lecture. In: Delage, P., Desrues, J., Frank, R., Puech, A. & Schlosser, F. (Eds.), *Proceedings of XVIII International Conference on Soil Mechanics and Geotechnical Engineering, Paris* (pp. 71–83).

Calabresi, G. & D'Agostino, S. (1997). Monuments and historic sites: Intervention techniques. In: Viggiani, C. (Ed.), *Proceedings of International Symposium on Geotechnical Engineering for the Preservation of Monuments and Historic Sites* (pp. 409–425). Rotterdam, The Netherlands: Balkema.

Como, M. (1993). Plastic and visco-plastic stability of leaning towers. In: *International Conference Physics Mathematics and Structural Engineering, in memory of Giulio Krall, Elba, Italy* (pp. 187–210).

De Silva, F., Ceroni, F., Sica, S., Pecce, M.R. & Silvestri, F. (2015). Effects of soil-foundation-structure interaction on the seismic behavior of monumental towers: The case study of the Carmine Bell Tower in Naples. *Rivista Italiana di Geotecnica*, 3, 7–27.

Desideri, A. & Viggiani, C. (1994). Some remarks on the stability of towers. In: Balasubramaniam, A.S. (Ed.), *Proceedings, Symposium on Development in Geotechnical Engineering (from Harvard to New Delhi, 1936–1994), 12–16 January 1994, Bangkok, Thailand*.

Federico, F. & Ferlisi, S. (1999). Time evolution of stability of leaning towers. In: Brebbia and Jager (Eds.), *6th International Conference STREMAH, Dresden* (pp. 485–494). WIT Press.

Gazetas, G. (1991). Foundation vibrations. In: Winterkorn, H.F. & Fang, H.-Y. (Eds.), *Foundation Engineering Handbook* (pp. 553–593). New York, NY: Van Nostrand Reinhold.

Hambly, E.C. (1985). Soil buckling and the leaning instability of tall structures. *The Structural Engineer*, 63(3), 77–85.

Heinle, E. & Leonhardt, F. (1988). *Türme: aller Zeiten, aller Kulturen* [*Towers: of all times, of all cultures*]. Stuttgart, Germany: Deutsche Verlags-Anstalt.

Heyman, J. (1992). Leaning towers. *Meccanica*, 27, 153–159.

Lancellotta, R. (1993). The stability of a rigid column with non-linear restraint. *Géotechnique*, 43(2), 331–332.

Lancellotta, R. (2013). La torre Ghirlandina: Una storia di interazione struttura-terreno. XI Croce Lecture. *Rivista Italiana di Geotecnica*, 2, 7–37.

Lancellotta, R. & Sabia, D. (2013). The role of monitoring and identification techniques on the preservation of historic towers. Keynote Lecture. *2nd International Symposium on Geotechnical Engineering for the Preservation of Monuments and Historic Sites, Napoli* (pp. 57–74).

Macchi, G. (1993). Monitoring medieval structures in Pavia. *Structural Engineering International*, 1, 6–9.

Marchi, M., Butterfield, R., Gottardi, G. & Lancellotta, R. (2011). Stability and strength analysis of leaning towers, *Géotechnique*, 61(12), 1069–1079.

Nova, R. & Montrasio, L. (1995). Un'analisi di stabilità del campanile di Pisa. *Rivista Italiana di Geotecnica*, 2, 83–93.

Pisanò, F., di Prisco, C.G. & Lancellotta, R. (2014). Soil-foundation modelling in laterally loaded historical towers. *Géotechnique*, 64(1), 1–15.

Geotechnics and Heritage: Historic Towers – Lancellotta, Flora & Viggiani
© 2018 Taylor & Francis Group, London, ISBN 978-1-138-03272-9

Tower of Pisa: Lessons learned by observation and analysis

M. Leoni
Wesi Geotecnica, Italy

N. Squeglia
University of Pisa, Italy

C. Viggiani
University of Napoli Federico II, Italy

ABSTRACT: The Leaning Tower of Pisa is one of the world's best known and most treasured monuments; it is founded on highly compressible soils and started leaning since the time of construction. In the 1990s the overhang had reached the value of 4.7 m and was increasing at a rate of 1.5 mm per year; a collapse appeared possible and even imminent, either by overturning or by brittle failure of the masonry. The International Committee appointed by the Italian Government in 1990 conceived and implemented a stabilisation intervention consisting in a small decrease of the inclination by underexcavation. Nowadays the stabilisation of the monument appears to have been attained, as confirmed by subsequent monitoring. A number of monographs and papers report the history of the monument and the progress of its inclination, the studies and investigations carried out since the early 20th century and the interventions of the Committee; these are shortly summarized at the beginning of the present chapter. A number of interventions carried out during the entire life span of the tower are discussed in detail, also taking advantage of advanced numerical analyses; it is shown that the Tower is very sensitive to any small disturbance and that some interventions, carried out with the best intentions, actually had definite detrimental effects. Possible future scenarios are finally briefly discussed.

1 A SHORT HISTORICAL SUMMARY

1.1 Introduction

The town of Pisa, located on the Arno River near the Tyrrhenian coast of Italy, became a flourishing commercial centre and a powerful maritime republic in the 11th century. Having defeated the Saracens in a series of sea battles it took control of the Mediterranean Sea and acquired colonies in Sardinia, Corsica, Elba, Southern Spain and North Africa. In the 12th century it was the naval base for the first Crusade and established a number of settlements in the Holy Land. The long-standing rivalry with Genoa finally culminated in the sea battle of Meloria (1284), where the Pisan fleet was destroyed; the power of the republic began to decline, and eventually, at the beginning of the 15th century, Pisa lost its independence and was annexed to the Republic of Florence.

It was in the period of maximum splendour of the Republic, in the 12th and 13th century, that the Cathedral, the Baptistery and the Bell Tower in Piazza dei Miracoli were erected (Figure 1). The Square is the awesome manifestation of the ideal unity that reigned at the time among religious, spiritual, and political powers; in its monuments civil history and history of art intertwine, giving them an extraordinary character as sign and symbol of the city (Franchi Vicerè *et al.*, 2005a, 2005b). Civic pride, identity, sense of belonging are evident in a stone on the façade of the Cathedral, recalling in an epic tone that construction was initially funded by the treasures captured from the Saracens after taking Palermo harbour in 1063.

The Leaning Tower (Figure 2) is one of the best known and most treasured monuments of the world; its extraordinary inclination turned it very early into a curiosity and a strong tourist attraction. The Tower consists

Figure 1. Pisa, Piazza dei Miracoli.

 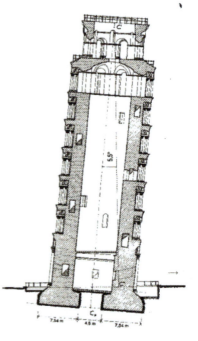

Figure 2. Pisa, the Leaning Tower.

of a hollow cylinder surrounded by six loggias with columns and vaults, merging from the base cylinder and surmounted by a belfry. The structure is thus subdivided into eight segments, called "orders".

The external surfaces are faced by a cut stone masonry; the annulus between the facings is filled with rubble and mortar within which extensive voids are found. A spiral staircase winds up within the annulus till the 6th order, while two shorter winding staircases lead to the floor and top of the belfry.

The staircase forms a large opening on the south side just above the level of the first cornice, where the cross section of the masonry suddenly decreases. The high stress within this region has been a major cause of concern, since it could give rise to an abrupt brittle failure of the masonry.

Records indicate that the Tower started leaning during construction (1173–1360); the movement continued over the centuries and in the 1990s the overhang had reached a value of 4.7 m, or an inclination of around 5.5°, and was increasing at a rate of 1.5 mm per year.

After the collapse of the S. Marco Bell Tower in Venice, in the early 20th century, the public opinion and the authorities turned their attention to the Tower of Pisa. The Italian government set up a number of Commissions that carried out extensive investigations; stabilisation measures were proposed, but no significant action was actually taken.

The sudden and unexpected collapse of the Civic Tower in Pavia (1989) drew new attention to the Leaning Tower. The Government closed the Tower to visitors in 1990 and set up an International Committee for the safeguarding of the Tower, chaired by prof. Michele Jamiolkowski, with the broad task of conceiving, designing and implementing the interventions needed to stabilize the monument.

The Committee focused first on gaining a comprehensive understanding of the previous studies on the monument and on elucidating the problems to be tackled. The respect of the formal, historic and material integrity of the Tower was recognized as a priority. A monitoring system was also installed and progressively extended and developed.

Being aware that the completion of their task would not have been a rapid matter, the Committee decided to implement some temporary and fully reversible interventions to improve the safety margin of the Tower with respect to overturning and to brittle failure of the masonry; these were carried out between 1992 and 1994.

The final intervention, as is well known, consisted in slightly decreasing the inclination of the Tower by underexcavation. It was carried out between 1998 and 2001 (Jamiolkowski, 2005), and brought back the inclination of the Tower to the value it had in the early 19th century.

On June 17, 2001, the Tower was returned to the authorities; since then, it has been continuously monitored by a Surveillance Group.

1.2 *The subsoil*

Figure 3 shows the soil profile underlying the Tower, consisting of three main horizons (Cestelli Guidi *et al.* 1971). Horizon A is about 10 m thick and consists primarily of estuarine deposits, laid down under tidal conditions; as a consequence, a rather erratic succession of sandy and clayey silt layers is found. At the bottom of horizon A there is a 2 m thick medium dense fine sand layer, the so-called Upper Sand.

Horizon B consists primarily of marine clay, which extends to a depth of about 40 m. It is subdivided into four distinct layers. The upper layer is a soft sensitive clay, locally known as the Pancone. It is underlain by an intermediate layer of stiffer clay, which in turn overlies a sand layer (the Intermediate Sand). The bottom layer of horizon B is a normally consolidated clay known as the Lower Clay. Horizon B is very uniform laterally in the vicinity of the tower.

Horizon C is a dense sand (the Lower Sand) which extends to considerable depth.

From the geological point of view (Trevisan 1971), the Lower Sands are marine sediments deposited during the Flandrian transgression. Horizon B consists of Quaternary marine clays deposited at the time of rapid eustatic rise. During the last 10,000 years or so the rate of eustatic rise decreased and the sediments became increasingly estuarine in character. The more recent sediments of horizon A comprise mainly sandy and clayey silt; as typical in estuarine deposits, there are significant variations over short horizontal distances. Based on sample descriptions and piezocone tests, it appears that the silty and clayey fraction south of the Tower are on

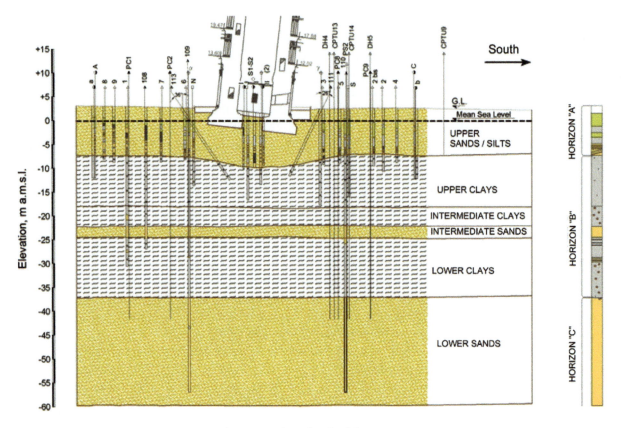

Figure 3. The subsoil of the Tower.

average larger than on its north, and the upper sand layer is locally thinner. This is believed to be at the origin of the southward inclination of the Tower.

The water table in horizon A is found at a depth of 0.5 m to 1.5 m below the ground surface; the latter has an average elevation of 2.5 m above mean sea level. Pumping from the lower sand results in downward seepage from horizon A with a pore pressure distribution with depth which is slightly less than hydrostatic.

The many boreholes beneath and around the Tower show that the surface of the Pancone clay is dished beneath the Tower, from which it can be deduced that the average settlement of the monument is not less than 3 m.

1.3 *History of construction and inclination*

Figure 4 reports schematically the history of construction of the Tower, as obtained from plenty of available documents. Works began in 1173 and construction had progressed to about one third of the way up the 4th order by 1178, when it was interrupted. The reason for this stoppage is not known, but had the construction continued much further, the foundations would have experienced a collapse (Burland, Potts, 1994).

The works were resumed in 1272, after a pause of nearly 100 years, by which time the strength of the ground had increased due to consolidation under the weight of the structure. By 1278, construction had reached the 7th cornice when it was interrupted again. As before, had the work continued, the tower would have fallen over.

The bell chamber was commenced around 1360 and completed around 1370, about two centuries after the start of the construction.

We know that the Tower must have been tilting right from the beginning of construction, because the ancient masons corrected for the inclination, giving finally to the Tower its so called "banana shape". The most important correction had to be applied to the belfry, due to the inclination developed during the second interruption of the works. On the north side, indeed, there are four steps from the seventh cornice up to the floor of the bell chamber, while on the south side there are six steps.

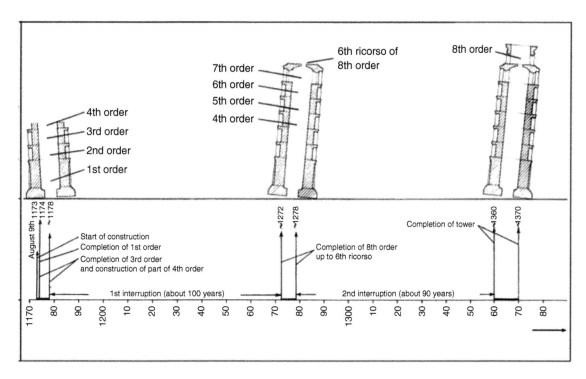

Figure 4. Schematic representation of the history of the construction.

Another important event in the history of the Tower was that in 1838 a walkway was excavated around its foundation. This, known as the "catino", had the purpose of exposing the column plinths and foundation steps for all to see as originally intended. The operation resulted in an inflow of water, because the bottom of the catino is well below the ground water table. As a consequence, since 1838, the catino had to be kept dry by continuous pumping.

In the belief that pumping could be dangerous for the stability of the Tower, between 1934 and 1935 its foundation and the soil surrounding the catino were made watertight by injecting cement grout into the foundation masonry and chemical grout into the soil. The intervention succeeded in effectively stopping the water inflow, and since then pumping was consequently interrupted.

A reliable clue on the history of the tilt lies in the adjustments made to the masonry layers during construction and in the resulting shape of the axis of the Tower. Based on this shape and on a hypothesis on the manner in which the masons corrected for the progressive lean of the tower, the history of inclination of the foundation of the tower reported in Figure 5 may be deduced. During the first phase of construction to just above the third cornice (1173 to 1178), the tower inclined slightly to the north. The construction stopped for almost a century, and when it recommenced in about 1272 the tower began to move towards the south. When the construction reached the seventh cornice in about 1278, the inclination was about 0.6° towards the south. During the next 90 years the construction was again interrupted and the inclination increased to about 1.6°, at an average rate of 40″/year (in terms of overhang, about 10 mm/year, i.e. much larger than the rate at the end of the 20th century). After the completion of the bell chamber in about 1370, the inclination went on increasing. Some information on its trend may be obtained by pictures or documents; among them, the value of the inclination deduced by a fresco painted in 1385 by Antonio Veneziano in the Camposanto and the value reported by Giorgio Vasari (1550). In 1817, when Cresy and Taylor (1829) made the first recorded measurement with a plumb line, the inclination of the tower was about 4.9°. Rohault de Fleury (1859) carried out another measurement 30 years later, finding a value of the inclination significantly higher than that by Cresy and Taylor. This may be explained by the fact that, between the two measurements, the catino had been excavated to uncover the base of the monument, which had sunk into the soil due to the 3 m settlement.

Since 1911 the inclination of the tower has been monitored by different means. The long term steady trend (Figure 6) is marked by two major perturbations: one in 1935 and another one in the early 1970s.

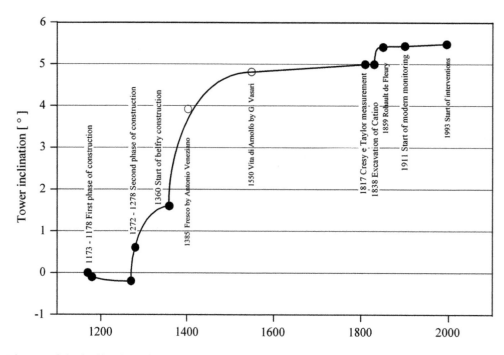

Figure 5. History of the inclination of the Tower. From 1173 to 1370 the inclination is deduced from the shape of the axis; from 1370 to 1817 from documents and pictures; since 1817, from direct measurements.

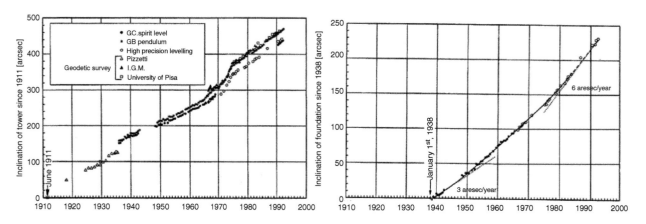

Figure 6. Measurements of the inclination since 1911. The diagram on the right is obtained by subtracting to the measured values the anomalous increments produced by different perturbations.

The first one was caused by the above mentioned cement grouting of the foundation body and the soil surrounding the catino, carried out to prevent the inflow of water. The second perturbation has been related to the pumping of water from deep aquifers, inducing subsidence all over the Pisa plain (Croce et al., 1981). The closure of a number of wells in the vicinity of the tower stopped the increase of the rate of tilt.

In any case, even correcting for the anomalous increments occurred in 1935, 1970–73 and some further minor perturbation, it appears that the rate of tilt was steadily increasing and had nearly doubled from 1938 to 1993. In the early 1990s the inclination was about 5.5° and increased at a rate of approximately 6 arcsec/year.

1.4 *The achievements of the International Committee, 1990–2001*

As recalled above, the International Committee for the Safeguard of the Tower of Pisa, chaired by a geotechnical engineer and formed by art historians, restorers, structural engineers and geotechnical engineers, was appointed in 1990 with the task of conceiving, designing and implementing the necessary stabilization works. The complete composition of the Committee is reported in Appendix 1.

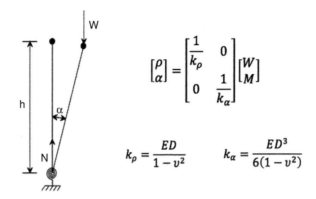

Figure 7. The inverted pendulum, a simple model for leaning instability.

A careful study of the behaviour of the tower led to the conclusion that it was affected by a phenomenon of instability of the equilibrium, known as leaning instability, depending on the stiffness and not on the strength of the foundation soil. A simple conceptual model of leaning instability is the inverted pendulum represented in Figure 7. The relationships between the vertical force W and the settlement ρ, and between the moment $M = We$ and the rotation α, can be obtained modelling the foundation as a rigid circular plate of diameter D resting on an elastic half space with Young's modulus E and Poisson ratio υ (Figure 7). In this simple linear model there is no coupling between settlement and rotation and the stability of the equilibrium is an intrinsic property of the ground-monument system, that may be expressed by the ratio FS between the stabilising and the overturning moment:

$$FS = \frac{k_\alpha \alpha}{Wh \sin \alpha} = \frac{ED^3}{6(1-\upsilon^2)} \frac{1}{Wh} \qquad (1)$$

We know that the settlement of the Tower of Pisa is $\rho \cong 3$ m, the weight of the Tower $W = 141.8$ MN, the height of the centre of gravity $h = 22.6$ m, the diameter of the foundation $D = 19.6$ m. It follows that $E/(1-\upsilon^2) \cong 2.41$ MN/m^2 and hence $FS \cong 0.95$. Even this simplistic linearly elastic subsoil model allows the important conclusion that the Tower is very nearly in neutral equilibrium.

In a case such as that of the leaning Tower, however, consideration of non linearity appears mandatory. The leaning instability of the Tower, therefore, has been investigated by a number of different approaches, including small scale physical tests at natural gravity and in the centrifuge and FEM analyses based on constitutive models of the soil accounting for non linearity, strain hardening plasticity and creep. One of the conclusions of these investigations was that a decrease of the inclination, even a relatively minor one, would result in a substantial increase in the stiffness of the foundation-soil system and, as a consequence, in a significant increase of the safety against leaning instability. This generated the idea that a decrease of the inclination could be used to stabilize the Tower.

As mentioned in the introduction, the Committee took an early decision to implement temporary interventions to improve the safety and hence to gain the time to properly devise, design and implement the permanent solution. Between May 1993 and February 1994 lead ingots for a total weight of 5.9 MN were placed on the north edge of the base of the Tower, producing a stabilising moment with a counter-rotation of 53″ and an additional settlement of 2.5 mm. The settlement and rotation produced by the counterweight were in good agreement with those predicted by the FE model, increasing the confidence in its reliability. The factor of safety FS increased from about 1 to 54; as a matter of facts the progressive southward inclination of the Tower came to a standstill and the Tower remained essentially motionless for over three years.

A medium term temporary scheme to replace the lead weights with ten tensioned steel cables anchored in the lower sands was developed for various reasons and attempted in 1995, but soon abandoned; we will come back on some aspects of this undertaking later on.

In the meantime, the Committee had come to the decision that the final stabilisation of the Tower should consist of decreasing its inclination by half a degree (1800 arcsec, i.e. around 10% of the inclination in 1990) as a final stabilisation measure, by inducing a proper differential settlement of the Tower. Acting on

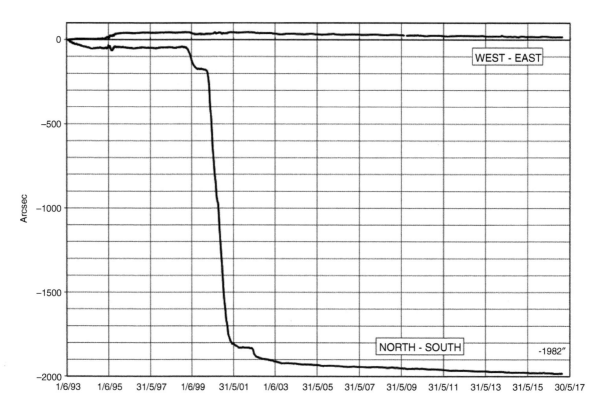

Figure 8. Rotation of the tower foundation since May 1993, evidencing the effects of preliminary and final underexcavation and of the drainage system.

the foundation soil and not on the tower, among other advantages such a solution is totally respectful of the formal, historic and material integrity of the monument. Three possible means to achieve the desired decrease of inclination have been investigated in detail, by extensive numerical modelling, small scale model tests at natural gravity and in the centrifuge and full scale field experiments, namely: (i) the construction of a ground pressing slab to the north of the tower; (ii) the consolidation of the Pancone clay north of the Tower by electro-osmosis, and (iii) the controlled removal of small volumes of soil beneath the north side of the foundation (underexcavation). The underexcavation was finally selected and implemented in two stages, achieving successfully the goal of reducing the inclination of 1800 arcsec (Figure 8) and contemporarily removing all the lead ingots.

Since the study of the movements of the Tower revealed that the oscillation of the ground water table consequent to heavy rainfall exerted a small but detectable negative influence on the inclination of the monument as a final intervention a drainage system aimed at stabilising the groundwater level in the vicinity of the Tower was installed in April 2002 (Figure 9). It produced a further reduction of the inclination of around 60″, clearly visible in Figure 8.

The Tower is now continuously monitored by a small Surveillance Group of experts formerly belonging to the International Committee; Figure 8 shows that at present the inclination is still decreasing at a very slow rate, which is approaching zero.

2 SOME LESSONS LEARNED BY OBSERVATION AND ANALYSIS

2.1 Introduction

In the following some interventions aimed at improving the stability or the fruition of the Tower, carried out during the history of the monument from the time of construction, will be discussed. It will be shown that some of them were successful and some detrimental; as it is well known, the way to hell is paved with good

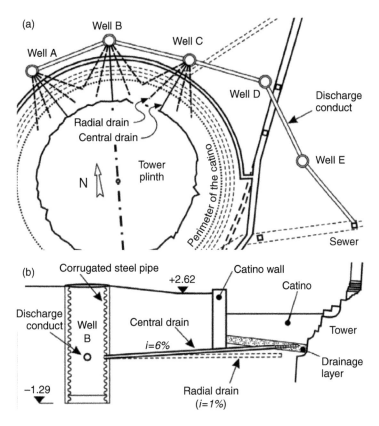

Figure 9. Drainage system to control the ground water table in Horizon A.

intentions. Hindsight shows that some of the detrimental effects of ill faced interventions might have been foreseen by the application of advanced methods of analysis.

2.2 *Shape of Tower axis and mass distribution*

An accurate survey of tower geometry (Locatelli *et al.*, 1971) revealed that the tower axis is not a straight line, but it is curved with the concavity northwards. Such a shape (Figure 10), currently known as the banana shape of the Tower, is due to the attempts of the ancient masons to compensate for the inclination resulting from differential settlement and consequent rotation of the foundation. The shape of the axis produces small offsets of the centre of gravity of the structure under construction; the resulting unbalanced moments have a great importance in terms of the present Tower inclination.

Leoni and Squeglia (2016) have recently developed a very sophisticated numerical model of the Tower and its subsoil, and used it successfully to reproduce the entire history of the monument. The Finite Element Mesh consists in 230000 tetrahedral elements and the Tower has been modelled as an elastic body with a heterogeneous density in order to reproduce the actual position of centroids of various orders. Main layers have been modeled by means of Soft Soil Creep model (Vermeer, Neher, 1999) and the soil parameters have been selected according to Potts (1993), Calabresi *et al.* (1996), and Mesri & Choi (1985). A special care has been dedicated to analyze the intervention carried out during the period 1993–2001 and a new numerical procedure has been introduced to simulate the effects of soil extraction or injection. This model has been used to analyse several scenarios. In particular, it has been used to analyse the difference between the rotation of the Tower foundation computed with the actual curved axis or in the assumption of no corrections, and hence with a straight axis (Figure 11). Surprisingly, the rotation of the tower without corrections is evaluated to be 4.3 degrees in 1993, which is considerably lower than the actual value of 5.5 degrees.

To explain this apparently strange result we have to remember that, in the first stage of the construction, the rotation of the tower had been northwards and hence the correction was southwards. During the second phase of construction the tower started to rotate southwards and the correction introduced since then aimed

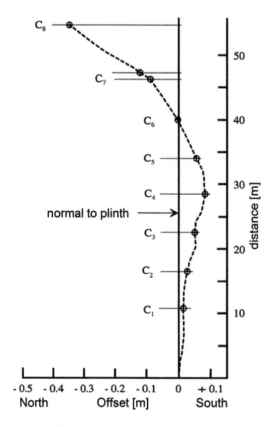

Figure 10. Shape of tower axis.

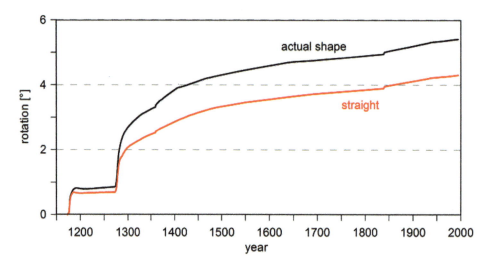

Figure 11. Computed influence of the shape of the tower axis.

to compensate the current overhang. The overall effect of these compensations consists in a centre of gravity of the whole tower slightly south to the normal to the foundation through its centre.

This is the first example (but not the last, as we will see later) of a detrimental measure implemented with the best intentions.

2.3 *Digging of the Catino*

In 1817 two English architects, students of Italian medieval architecture, carried out a survey of the Pisan monuments including the Leaning Tower. Some years later they published a detailed cross section of the Tower

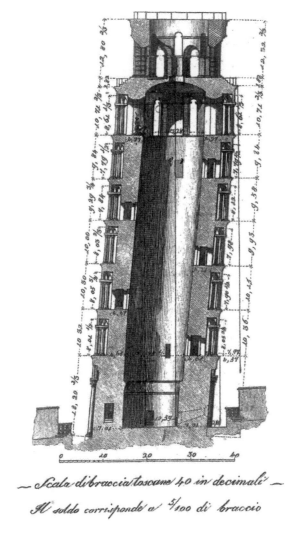

Figure 12. The survey carried out in 1817 by Cresy and Taylor (1829). The positions of the plumb lines used to measure the inclination are reported.

(Figure 12) reporting the locations of plumb lines they had used to measure the inclination (Cresy, Taylor, 1829); it was thus possible to establish a definite correspondence with the later measurements.

Figure 5 suggests that, in the early XIX century, the Tower was motionless or, at most, it was increasing its lean at a very small rate.

Forty years later, however, the French architect Rohault de Fleury (1859) carried out new measurements and found a value of the inclination significantly larger than that of Cresy and Taylor (5.4° instead of 5°). This apparent discrepancy gave rise to heated discussions, and was attributed to an error in one of the surveys; nobody could even conceive that the Tower had actually increased its inclination.

As a matter of facts in 1835, and hence between the two surveys, the Pisan architect Alessandro Gherardesca had excavated the Catino around the base of the Tower. This intervention was intended to uncover the base of the monument which had sunk into the soil due to the 3 m settlement; Gheradesca (1835) proudly claimed that he was moved by love of his homeland and by the desire of saving the vestiges of the ancestral greatness, and regretted that the ancient people had only been able to fill the settlement trough with soil.

Digging the catino, however, caused an increase of inclination of approximately 0.5°; more importantly, the rate of inclination increased and the motion changed from retarded to accelerated. Nowadays, the negative effect of excavating around the base of a tower is evident on the basis of simple bearing capacity considerations; and filling with soil appears to be a rather rational intervention. It is somewhat surprising that nobody was afraid of this danger; probably the stability of an ancient and somewhat mythic Tower was taken for granted.

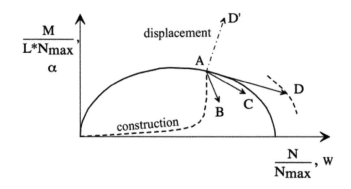

Figure 13. Yield locus and load increments due to the counterweight.

On the contrary, the inflow of water into the catino and the related transport of fines has been always considered a significant risk and the increase of inclination of Tower was attributed to fines transport. As recalled above, to eliminate this danger, in the years 1934–35 the Tower foundation and the soil surrounding the catino were made watertight by injecting cement grout into the foundation masonry and chemicals into the soil. The intervention succeeded in stopping the water inflow, and therefore pumping was interrupted; during the intervention, however, some excavation around the base of the tower had to be carried out again and an anomalous increase of the inclination occurred.

The excavation of the Catino is another important example of a detrimental intervention carried out with the best intention. The back analysis of the history of the Tower carried out by Leoni and Squeglia (2016) confirms that, had advanced FEM been available to Gherardesca in the mid XIX century, he might have foreseen that digging the catino would have resulted in an increase of the inclination and of the rate of inclination.

2.4 *The lead counterweights*

As mentioned above, as the Tower was on the verge of leaning instability, the Committee took an early decision to implement temporary interventions to improve the safety and, hence, to gain the time to properly devise, design and implement the permanent solution. From a structural point of view, a hoop of pre-tensioned steel wires was installed around the first cornice. From a geotechnical point of view, between May 1993 and February 1994, 64 lead ingots were installed on the north edge of the base of the Tower for a total weight of 5.9 MN, producing a stabilising moment of about 41 MNm with a counter-rotation of $53''$ and an additional settlement of 2.5 mm.

It must be pointed out that the counterweight induces a stabilising moment, but also an increase of the vertical load applied to the foundation. The non-linear analysis of leaning instability, however, had demonstrated that there is a coupling between rotation and settlement, so that an increase of the vertical load might result in a southward rotation. This dangerous perspective required that a detailed analysis should be carried out before implementing the intervention.

The behaviour of a shallow foundation subjected to a vertical force N and a moment M has been investigated by many authors (Nova, Montrasio, 1991; Dean *et al.*, 1992; Desideri, Viggiani, 1994; Nova, Montrasio, 1995; Desideri *et al.*, 1997). It may be described using a yield locus in the plane N, M, similar to that reported in Figure 13. Inside the yield locus an infinity of similar loci may be imagined, expanding by hardening with increasing loads. Superposing to the axis N, M the axis of settlement ρ and rotation α, a qualitative description of the increments of displacement following a load increment may be obtained.

The point representing the conditions of the Tower in 1993 (point A in Figure 13) is located on a yield locus, since it is the end point of a monotonic loading process consisting first in an increase of N (construction) followed by an increase of M (progressive inclination). Starting in A any load increment directed inside the locus, such as that represented by vector AB, produces an elastic displacement. The increment of displacement is coaxial with the increment of load, and thus a decrease in moment M produces a decrease of the rotation α, irrespective of the value of vertical load.

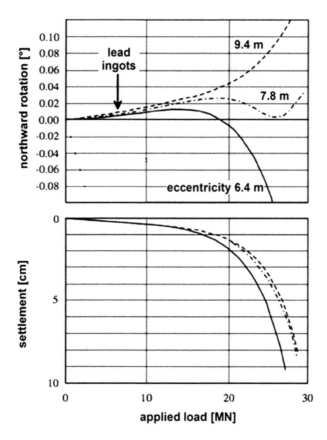

Figure 14. Effects of counterweights application in terms of tower rotation and settlement with applied load in dependence of eccentricity.

On the contrary, if the load increment is directed outwards, such as that represented by the vector AD in Figure 13, elasto-plastic displacements occur and the locus expands by hardening. The plastic displacement increment, that is almost the totality of the displacements, is not coaxial with the load increment but directed normally to a plastic potential. For a material with an associated law of flux the displacement increment is normal to the yield locus, for example as the vector AD′ in the figure. A decrease of the moment might thus result in an increase of the rotation.

FEM analyses (Burland, Potts, 1994) demonstrated that the effect of the counterweight depends on the position of the force; the greater the lever arm, the greater the effect in terms of reduction of the inclination. On the contrary, for a small lever arm or for a large applied load, a negative effect of increase of the inclination may occur.

Figure 14 shows some predictions of the rotation induced by the counterweight with varying eccentricity and applied load. Since the actual eccentricity of the counterweights was about 7 m, the maximum applied load had to be limited and there was the risk of a negative effect; a comprehensive monitoring was carried out during counterweight application to detect and control any detrimental effect. The 5.9 MPa actually produced a northwards rotation by 53 arcseconds with a settlement by 2.5 millimetres (Figure 15): a positive effect obtained and a risk avoided thanks to the understanding achieved by means of numerical analysis.

2.5 The 10 anchors solution

The safety of the Tower had been successfully improved, on a temporary basis, by the lead ingots counterweight. While continuing to study a final stabilisation measure, the Committee entered a period of uncertainty about the real possibility of concluding their work; for a number of reasons, there was a widespread fear that the Committee may dissolve, as all the previous Committees had done, but this time with the aggravating factor of leaving the Tower spoiled by the stack of lead ingots for a period of time difficult to foresee and certainly not short.

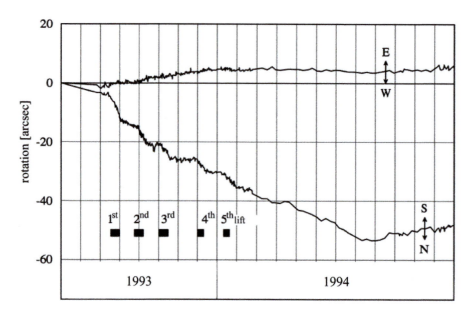

Figure 15. The decrease of inclination produced by the application of lead counterweight.

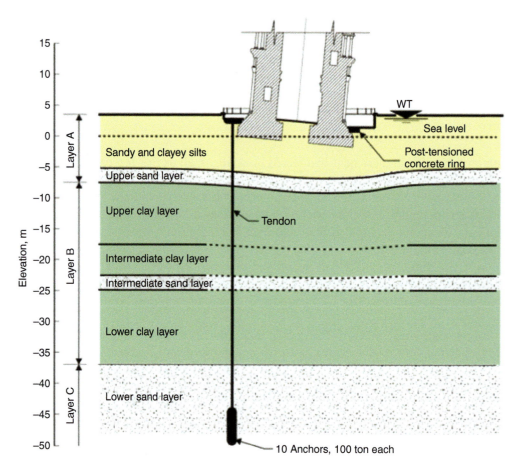

Figure 16. The medium term solution with ten ground anchors.

Essentially for this reason, a medium term temporary scheme was devised to replace the lead weights with ten tensioned steel cables anchored in the Lower Sands at a depth of more than 40 m (Figure 16). Apart from the advantage of being invisible, an additional benefit of this scheme was the possibility to increase the lever arm that thus obtaining a stabilizing moment larger than that of the lead ingots. The main problem of the ten

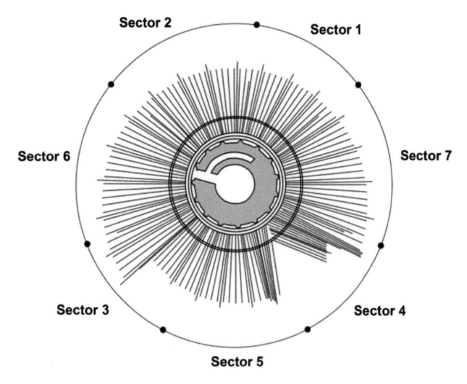

Figure 17. Arrangement in plan of freezing tubes.

anchors solution was that the anchors had to be connected to the Tower foundation through a ring beam to be constructed below the floor of the catino, and this involved carrying out an excavation around the Tower below the ground water level, an operation of the utmost delicacy; the experience of the excavation of the Catino in the mid XIX century was there to warn against the risks involved. After a careful comparison of different options, it was decided to employ local ground freezing immediately below the catino floor but well above the Tower foundation level. The post tensioned concrete ring was to be installed in short sections so as to limit the length of excavation open at any time. Before commencing the operations, exploratory drilling through the floor of the catino revealed the existence, below the catino floor, of a 1 m thick concrete bed, cast in part in 1837 and in part in 1935. A well defined gap at the interface between the concrete and the tower foundation led to the conclusion that the two were not connected; as a consequence, the volume variations of the frozen soil during freezing and thawing were not expected to affect the Tower.

Freezing of the soil around the catino and the foundation of the Tower was performed by circulating liquid nitrogen into 157 freezing tubes installed in inclined boreholes (Figure 17).

The concrete bed was cut with a wire saw in sectors, and the corresponding sectors of the ring beam were immediately cast in situ (Figure 18).

Freezing started on the north side of the Tower and the northern sections of the ring were successfully installed between May and July 1995. The freezing operations consisted of 36 hours during which liquid nitrogen was continuously circulated, followed by a maintenance phase when freezing was carried out for one hour per day, so as to limit the expansion of the ice front. Some warrying southward rotation of the Tower did take place during freezing on the north side, but this was recovered on thawing (Figure 19); some lead ingots were added on the stack to control the southward motion. In mid July, summing the effects of lead weights, installation of the freezing tubes and construction of the two northern sectors of the ring beam, the Tower had experienced a northward rotation of 60″ and a mean settlement of 4.4 mm.

In September 1995 freezing was commenced on the south-west and south-east sides of the foundation; in August, as a prudential measure, 29 micropiles had been installed through the wall of the catino on the south side. As a consequence of the freezing operations and of the micropiles installation the Tower underwent significant southward rotations. The operation was suspended and the rotation controlled by adding further lead ingots to the stack and by decreasing the level of the water table north of the Tower by pumping.

Figure 18. The construction of the north sectors of the ring beam.

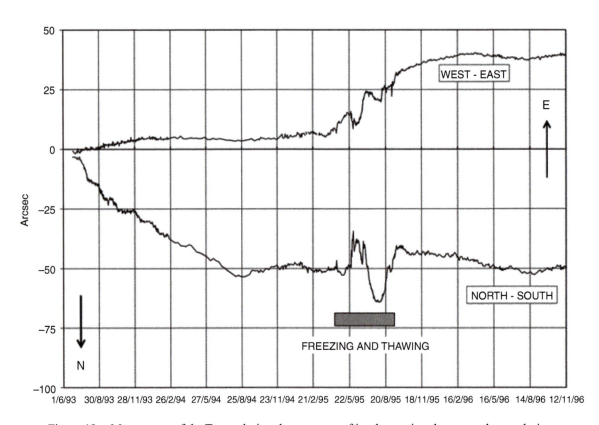

Figure 19. Movements of the Tower during the attempts of implementing the ten anchors solution.

At that time it was discovered that a large number of steel grout-filled pipes connected the conglomerate to the masonry foundation. These had been installed in 1935 when the foundation masonry was grouted, and the Committee was not aware of their existence. In view also of the uncertainty about the strength of this connection, the whole ground anchors intervention was abandoned.

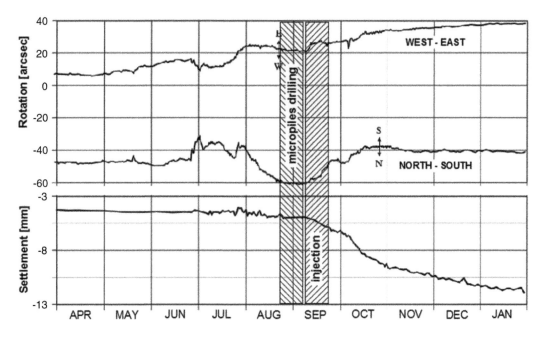

Figure 20. Effects of micropiles installation on the movements of the tower.

The overall southward rotation of the Tower had been relatively small (about 7″ in comparison to the about 60″ gained by the application of the counterweights).

A detailed scrutiny of the tower movements during the installation of the micropiles is reported in Figure 20.

It appears that the installation of micropiles played the main role in the undesired tower movements. As a matter of fact, a numerical analysis (Leoni, Squeglia, 2016) confirmed that the injection phase of micropiles on the south side of the Tower induces an increase of inclination. Again, should the advanced numerical model have been available twenty years ago, the risk of detrimental effects may have been foreseen.

2.6 Underexcavation

As it is largely known, the Leaning Tower has been eventually stabilised by means of a reduction of its inclination; the physical basis of the measure has been explained in § 1.4 above and additional information can be found elsewhere (Viggiani et al., 2005).

After careful consideration of a number of possible methods to reduce the inclination, the Committee chose to study three of them in detail, namely: (i) a ground pressing slab to the north of the Tower, to be connected structurally to the foundation of the Tower; (ii) electroosmotic consolidation of the Upper Clay beneath the north side of the foundation; (iii) the controlled removal of small volumes of soil beneath the north side of the foundation (underexcavation).

All three measures could produce negative effects if applied in an inappropriate way. In particular, numerical analysis and centrifuge modelling of the pressing slab showed that the response of the Tower was uncertain and, if positive, not sufficient to reach the desired effect; furthermore, the connection of the slab with the foundation would be very difficult. Laboratory investigations and a preliminary analysis of the electroosmotic solution were encouraging; a full scale field trial was thus undertaken (Viggiani, Squeglia, 2003). Unfortunately, it proved to be a total failure, as a number of potentially dangerous unexplained effects were observed, electro osmosis had thus to be ruled out as a suitable method to stabilize the Tower.

Underexcavation was modelled by small scale physical testing at 1g and in the centrifuge; the effectiveness of the method was evident, provided that soil extraction took place beyond a "critical line" located about half a radius north the centre of the foundation. Numerical analyses confirmed this important feature, that was unexpected and that imposed a well defined limit to the intervention, to prevent any negative effect.

The results of physical and numerical modelling encouraged the undertaking of a field trial, that was carried out (Figure 21) in autumn 1995 on a 7 m diameter plinth eccentrically loaded and extensively instrumented.

Figure 21. Trial field of underexcavation: view of plinth and drilling machine.

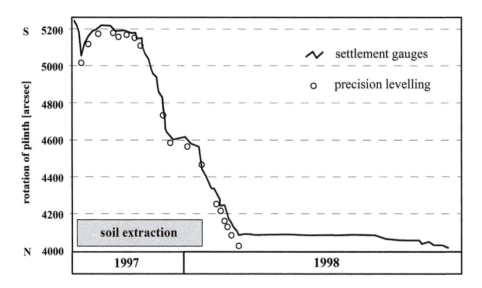

Figure 22. Underexcavation of the trial plinth.

The objectives of the trial were: (i) to evaluate the effectiveness of the method; (ii) to develop a suitable excavation technique; (iii) to measure the changes in contact stress and pore water pressure beneath the foundation during the excavation process; (iv) to explore methods to "steer" the tower during the process by adjusting the drilling sequence.

The trial was very successful. The plinth was rotated by about 900 arcsec maintaining a satisfactory directional control (Figure 22); the stress changes beneath the foundation were rather small (Figure 23); the rotational response was rapid and the plinth came to rest rapidly on completion of underexcavation.

It is important to note that, early in the trial, overenthusiastic drilling resulted in soil extraction beyond the critical line causing a counter rotation; the trial therefore confirmed again the existence of the critical line.

The final underexcavation beneath the Tower, obviously, took place on the right side of the critical line and was the means of stabilization finally successfully employed.

The small and rather obvious lesson that can be obtained is that a careful study of any intervention is a guarantee of success even for difficult problems.

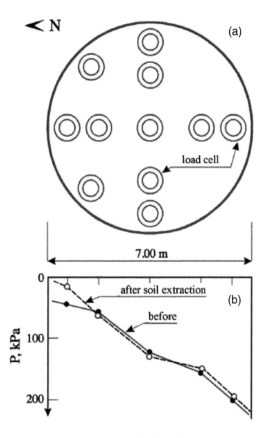

Figure 23. Contact stress beneath the plinth before and after underexcavation.

3 GOOD LUCK

Most major undertakings rest, at least partially, on the effect of casual favourable circumstances, and the Leaning Tower is not an exception. There are at least three instances in which the Tower has been saved by unbelievably fortunate circumstances.

As already recalled in § 1.3 above, in 1178 and 1278 the construction of the Tower had been interrupted for two rest periods of about one century each. It can be shown that, had the construction continued much further, the tower would have experienced a collapse. This circumstance has been firstly put into evidence by Burland and Potts (1994) by means of a numerical analysis based on an equivalent 2D model. Further analyses have been carried out by means of the advanced 3D numerical model developed by Leoni and Squeglia (2016). Figure 24 reports the results: the actual history of the inclination of the Tower is compared with that obtained in the hypothesis of a construction without interruptions. It may be seen that the inclination would have increased rapidly up to a value over 5.5°; at such an inclination, the stress in the masonry of the Tower is equal to zero on the north side and near 14 MPa on the south side. The Tower would have collapsed either by structural failure or by overturning approaching an inclination above 6°.

Modern geotechnical engineers have been tempted by the idea that the ancient masons had an understanding of soil mechanics and implemented a staged construction to take advantage of pore pressure dissipation and increased rigidity of the soil; but no traces of even a minimum worry about the statics of the Tower can be found in the very rich documentation available in the archives about the construction of the monument. One cannot but conclude that fate was at play!

In July 1944, during the Second World War, Pisa was divided: Germans were resisting at the north, while Americans advanced from the south. Every bend of road, every farmhouse and every escarpment seemed to be occupied by obstinate German defenders. As the number of deceased and wounded Americans mounted, the advance was in danger of stalling. How could the German be so accurate in such a flat, coastal terrain? They had to have a vantage point; may it be the Leaning Tower?

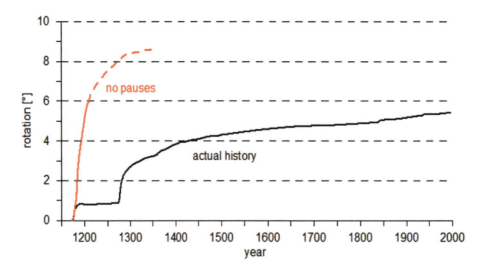

Figure 24. Comparison between the actual history of rotation and the rotation consequent to a construction without interruptions.

Figure 25. Sergeant Leon Weckstein in 1944 (left) and at the time of writing his book.

Sergeant Leon Weckstein (Figure 25) was entrusted the most important mission of his war: to get close to the Tower to find out if the Germans were inside. If any enemy activity had been detected, Americans were not going to sacrifice men for a chunk of masonry, no matter how old.

"I took my time," Weckstein (1999) writes in his book "training the binocular slowly up and down, attempting to discern anything that might be hidden within those recess and arches". But after a whole day of observation he did not call down fire. Waiting for the signal there were inland gun batteries and a destroyer offshore.

What the 91st Infantry Division did not know was that they were entrusting one of the war's most fateful missions to a man rejected by the navy for being short-sighted. "In 1942 the navy had told me to go away and eat carrots for six months," says Weckstein. "Then the infantry took me – but they take anyone"!

It is thus very probable that the Tower survived the war thanks to the short-sightedness of an American officer.

4 CONCLUDING REMARKS

The long and tormented history of the Leaning Tower, from its construction until the attempts to stabilize it, is rich in lessons.

Firstly, the mastery of the ancient Pisans and their perseverance in completing the construction over a time span of two centuries and in spite of the occurrence of an evident inclination, and of a number of political and economic difficulties, find a match in the obstinacy of modern engineers who succeeded in saving the Tower for the admiration of future generations through over a century of attempts and in spite of many difficulties and some errors. First lesson: perseverate!

Another lesson is that good intentions are not enough to ensure success; it is proverbial to say that they pave the way to hell. This is confirmed by the corrections carried out during construction to counteract the inclination, that finally proved detrimental; by the excavation of the catino, which brought the Tower on the verge of collapse; by the attempt of excavating a ring beam below the catino floor for the ten anchors solution, and particularly by the installation of micropiles, which increased the inclination of the Tower. Advanced numerical analyses, carried out after the event, help to explain the behaviour observed; this means that, in principle, if used in advance they might have avoided the risks.

A final lesson, very important for the two Neapolitan authors, is that a modicum of good luck won't do any harm!

Fifteen years after the end of underexcavation it can be agreed that the intervention has been successful. At present the Tower is still moving northward at a rate of rotation of $4''$ per year.

Two possible scenarios have been depicted for the future. In the optimistic one the progressive increase of the inclination, affecting the tower from the time of construction, has been stopped and the monument will keep motionless, apart from the cyclic movements connected to the environmental actions, such as the daily sun irradiation and the seasonal groundwater table fluctuations.

The pessimistic scenario sees the Tower staying motionless for a period of some decades (the honeymoon), followed by a resumption of the southward rotation and approaching again the value it had experienced at the end of the 20th century.

At present, the results of monitoring seem to support the optimistic scenario. On the contrary, the finite element analyses indicate that the pessimistic scenario might be the right one. In the next – probably few – years, the Tower itself will provide the answer to our questions. In any case, should the inclination become again dangerous in the future, the underexcavation intervention could be repeated!

ACKNOWLEDGEMENTS

The content of this chapter is largely based on the work of the International Committee and of the subsequent Surveillance Group appointed with the task of supervising and interpreting the monitoring; the composition of the Committee is reported in Appendix 1.

In particular, the authors wish to acknowledge the substantial contributions of Michele Jamiolkowski, John B. Burland and Salvatore Settis.

REFERENCES

Burland, I.B., Potts, D.M. (1994) Development and application of a numerical model for the leaning tower of Pisa. *IS Prefailure Deformation Characteristics of Geomaterials,* Hokkaido, Japan, vol. 2, 715–738.

Calabresi, G., Rampello, S., Callisto, L., Viggiani, G.M.B. (1996) The Leaning Tower of Pisa – Soil parameters for the numerical modelling of the tower resulting from the most recent investigations. Università di Roma "La Sapienza": Laboratorio Geotecnico Dipartimento di Ingegneria Strutturale e Geotecnica.

Cestelli Guidi, C., Croce, A., Skempton, A.W., Schultze, E., Calabresi, G., Viggiani, C. (1971) Caratteristiche geotecniche del sottosuolo della Torre. *Ricerche e studi sulla Torre pendente di Pisa ed i fenomeni connessi alle condizioni d'ambiente,* IGM, Firenze, I, pp. 179–200.

Cresy, E., Taylor, G.L. (1829) Architecture of the Middle Ages in Italy: illustrated by views, plans, elevations, sections and details of the cathedral, baptistry, leaning Tower of campanile and campo santo at Pisa from drawings and measurements taken in the year 1817, London: published by the Authors.

Croce, A., Burghignoli, A., Calabresi, G., Evangelista, A., Viggiani, C. (1981) The Tower of Pisa and the surrounding square: recent observations, *X International Conference on Soil Mechanics and Foundation Engineering,* Stockholm, III, pp. 61–70.

Dean, E.T.R., James, R.G., Schofield, A.N., Tan, F.S.C., Tsukamoto, Y. (1992) The bearing capacity of conical footings on sand in relation to the behaviour of spudcan footings of jackups. Proc. Wroth Mem. Symp. on Predictive Soil Mechanics, Thomas Telford, London, pp. 230–253.

Desideri, A., Viggiani, C. (1994) Some remarks on the stability of towers, *Symposium on development in Geotechnical Engineering, from Harvard to New Delhi 1936–1994*, A.I.T. Bangkok.

Desideri, A., Russo, G., Viggiani, C. (1997) Stability of towers on compressible ground, *Rivista Italiana di Geotecnica*, 31(1): pp. 5–29.

Franchi Vicerè, L., Viggiani, C., Squeglia, N. (2005a) *La Piazza del Duomo: sottosuolo, archeologia, storia* in La Torre restituita. Gli studi e gli interventi che hanno consentito la stabilizzazione della Torre di Pisa, Volume Speciale del Bollettino d'Arte del Ministero per i Beni e le Attività Culturali, Roma.

Franchi Vicerè, L., Veniale, F., Lodola, S., Pepe, M. (2005b) *La Torre campanaria* in La Torre restituita. Gli studi e gli interventi che hanno consentito la stabilizzazione della Torre di Pisa, Volume Speciale del Bollettino d'Arte del Ministero per i Beni e le Attività Culturali, Roma.

Gherardesca, A. (1835) Considerazioni sulla pendenza del Campanile della Primaziale Pisana. *Pisa*.

Jamiolkowski, M. B. (2005) *Introduzione* in La Torre restituita. Gli studi e gli interventi che hanno consentito la stabilizzazione della Torre di Pisa, Volume Speciale del Bollettino d'Arte del Ministero per i Beni e le Attività Culturali, Roma.

Leoni, M., Squeglia, N. (2016) *3D Creep analysis of the leaning tower of Pisa*. Opera della Primaziale Pisana. Scientific report.

Locatelli, P., Polvani, G., Selleri, F. (1971) Caratteristiche geometriche e fisiche della Torre e suo stato di conservazione, *Ricerche e studi sulla Torre pendente di Pisa ed i fenomeni connessi alle condizioni d'ambiente*, IGM, Firenze, I, pp. 17–68.

Mesri, G. & Choi, Y.K. (1985) The uniqueness of the end-of-primary (EOP) void ratio-effective stress relationship. In Proc. 11th ICSMFE. San Francisco, 1985.

Nova, R., Montrasio, L. (1991) Settlement of shallow foundation on sand. *Geotechnique*, 41, n. 2, pp. 243–256.

Nova, R., Montrasio, L. (1995) Un'analisi di stabilità del Campanile di Pisa. *Rivista Italiana di Geotecnica*, vol. XXIX, n. 2.

Potts, D.M. (1993) Calibrazione di un modello geotecnico agli elementi finiti e valutazione degli effetti indotti a seguito di alcuni interventi di consolidamento della Torre di Pisa. Affidamento n. JAM 3018.35/tp.

Rohault de Fleury (1859) Le Campanile de Pise, *Encyclopedie de l'Architecture*, Paris: Bance.

Trevisan, L. (1971) Caratteri geologici, chimici e mineralogici del sottosuolo della Torre e nei pressi di essa, *Ricerche e studi sulla Torre pendente di Pisa ed i fenomeni connessi alle condizioni d'ambiente*, IGM, Firenze, I, pp. 151–164.

Vasari, G. (1550) Le vite de' più eccellenti architetti, pittori et scultori italiani da Cimabue infino a' tempi nostri, *Lorenzo Torrentino, Firenze*.

Vermeer, P.A., Neher, H.P. (1999) A soft soil model that accounts for creep. In Proc. Int. Symp. "Beyond 2000 in Computational Geotechnics". Amsterdam, 1999. Balkema.

Viggiani, C., Squeglia, N. (2003) Electroosmosis to stabilize the leaning tower of Pisa, *Rivista Italiana di Geotecnica* 37(1): 29–37.

Viggiani, C., Squeglia, N., Pepe, M. (2005) *Studio dei metodi per la stabilizzazione della Torre* in La Torre restituita. Gli studi e gli interventi che hanno consentito la stabilizzazione della Torre di Pisa, Volume Speciale del Bollettino d'Arte del Ministero per i Beni e le Attività Culturali, Roma.

Weckstein, L. (1999) Through my eyes; the 91st Infantry Division in the Italian Campaign 1942–1945. Hellgate Press, ISBN 1-55571-497-8.

APPENDIX 1

The International Committee for the safeguard of the Leaning Tower of Pisa was set up by the Prime Minister in March 1990, with the task of conceiving, designing and implementing the interventions needed to stabilise the Tower. It was conceived from the very beginning as a truly interdisciplinary body.

In June 2001, when the Tower was given back to the City of Pisa after its stabilisation, the composition of the Committee was as follows.

For Art History, Restoration and Construction Materials: Jean Barthelemy, Michele D'Elia, Roberto Di Stefano, Alma Maria Mignosi, Salvatore Settis and Fernando Veniale.

For Structural Engineering: Remo Calzona, Giuseppe Creazza, Giorgio Croci, Giorgio Macchi and Luca Sanpaolesi.

For Geotechnical Engineering: John B. Burland, Michele Jamiolkowski (Chairman) and Carlo Viggiani.

During its eleven years of activity the Committee included other scientists and experts: for Art History and Restoration, Michele Cordaro, Francesco Gurrieri, Raymond Lemaire and Angiola Maria Romanini; for Structural Engineering, Mario Desideri and Fritz Leonhardt; for Geotechnical Engineering, Renato Lancellotta and Gerald A. Leonards.

Quite a number of other scientists and experts from Italy and abroad cooperated with the Committee on different problems with great enthusiasm and willingness; among them, Raffaello Bartelletti, Giovanni Calabresi, Gisella Capponi, Ezio Faccioli, Giuseppe Grandori, James K. Mitchell, David Potts, Giovanni Solari.

The Committee was supported by a Consortium including an Engineering Company (BONIFICA S.p.A., Rome), three Foundations Contractors (ITALSONDA S.p.A., Naples; RODIO S.p.A., Milan; TREVI S.p.A., Cesena) and an Applied Research Institute (ISMES, Bergamo).

It is hoped that the fruitful methodological approach developed for the Tower of Pisa might embody helpful indications for other restoration interventions throughout the world in the future.

Geotechnics and Heritage: Historic Towers – Lancellotta, Flora & Viggiani
© 2018 Taylor & Francis Group, London, ISBN 978-1-138-03272-9

The Ghirlandina Tower in Modena: An example of dynamic identification

R. Lancellotta & D. Sabia
Department of Structural, Geotechnical and Building Engineering, Politecnico di Torino, Italy

ABSTRACT: Within the context of the restoration principles, that claims for preserving the whole integrity of the monument, it should be well accepted that we should not rush in deciding the stabilizing measures until the behavior of the monument is properly understood.

This requires a long term monitoring aimed at identifying the causes of movements, their potential increase in rate and requires to distinguish the normal physiological behavior from deviations.

In addition, these observations can be complemented with experimental identification analyses, performed in the presence of ambient vibration, that allow to explore in a rather effective way the response of the structure and its interaction with the soil.

This chapter is intended to contribute on the subject of preservation of historic towers by highlighting, throughout the well documented case history of the Ghirlandina tower, the role of continuous dynamic monitoring and identification techniques for an accurate assessment of the response of the structure, including its interaction with the foundation soil.

1 INTRODUCTION

Historic masonry towers are an important part of our cultural heritage. Some of them originated as bell towers, and among the oldest examples there are the campaniles in Ravenna, well known for their peculiar cylindrical shape and for not being structurally connected to the church. During the 10th and 11th centuries there was an impressive development of private towers (this was the case of Bologna, S. Gimignano, Florence, Lucca, Pisa, Pavia and many other cities), used by families as their house in case of strife within the medieval commune and as a symbol of their power and prosperity. In the thirteenth century the landscape of the commune was being dominated by the civic tower and the bell tower of the cathedral. Some others were defense tower, as it was the case of the fortified town of Monteriggioni (Siena) with fourteen towers at the corner of the defence wall. All these example, and many others, tell us that historic masonry towers are part of our history and our culture, i.e. they are not only a distinctive feature of old towns but, today as in the past, they represent the local community as a whole. This is because they are a part that stands for the whole and generate deep consciousness of belonging to a community. For all these mentioned aspects their preservation deserves special attention and efforts.

It seems evident that the towers today we observe survived the initial period in which they could have been not so far from a *bearing capacity* collapse, due to *lack of strength* of the soil. Delay or interruption of the building process allowed the foundation soil to improve its strength and the tower to be successfully finished. And due to uneven settlements, they appear today in some circumstances to survive at an alarming angle of inclination. This highlights the danger of a *leaning instability*, especially in the case of slender towers, that increases if there is *lack of stiffness* of the soil. In this respect the leaning tower of Pisa represents a powerful example, because the preservation of the tower was recognised as being a problem of leaning instability (Hambly, 1985; Lancellotta, 1993; Como, 1993; Desideri and Viggiani, 1994; Nova and Montrasio, 1995; Federico and Ferlisi, 1999; Burland *et al.*, 2003; Marchi *et al.*, 2011).

Obviously, there are also problems of structural nature interacting with these geotechnical aspects. They depend on the masonry behaviour as a unilateral material and deserve special attention as proved by the collapse of the Campanile in Venice in 1902 and the Civic Tower in Pavia in 1989 (Heyman, 1992; Binda *et al.*, 1992; Macchi, 1993).

A further key aspect for the preservation is related to their seismic vulnerability, as unfortunately confirmed by the Emilia earthquake in May 2012 and the more recent Centre Italy earthquake of August–October 2016, during which many historical masonry *campaniles* (bell towers) collapsed.

By considering all the mentioned examples, two strictly related aspects merge: it is evident how the preservation of historic towers deserves special attention, but at the same time, it must be stressed that there are important constraints when devising remedial measures and intervention techniques, due to the need of preserving the integrity of the monument. It has to be recalled that quite often the integrity requirement is only interpreted as the requirement of preserving the shape and the appearance of the monument. In reality it also implies historic integrity, by considering the changes the monument experienced with time, as well as material integrity, that means construction techniques, materials and structural scheme. Therefore, preserving integrity requires a multidisciplinary approach as well as to develop the attitude not to rush in deciding the stabilizing measures until the behavior of the monument is properly understood (Brandi, 1977; Calabresi and D'Agostino, 1997; Calabresi, 2013).

This chapter is intended to contribute on the subject of preservation of historic towers by highlighting, throughout the well documented case history of the Ghirlandina tower, the role of continuous dynamic monitoring and identification techniques for an accurate assessment of the response of the structure. It also focuses on the influence of the soil-structure interaction on such a response.

2 THE CASE HISTORY OF GHIRLANDINA TOWER

The Modena Cathedral and its bell tower, the Ghirlandina (Figure 1), are cornerstones of the Italian cultural heritage (Brandi, 1985) and part of the Piazza Grande in Modena (Italy), Unesco site.

Figure 1. The Cathedral of Modena and its Ghirlandina Tower.

The construction of the Cathedral was realized between 1099 and 1319, when the construction of the Ghirlandina Tower was also completed. Inscriptions on the façade and on the central apse celebrate the sculptor Wiligelmo and the architect Lanfranco, respectively. Approximately 25 m wide and 66 m long, with a roof height of roughly 24 m, the Cathedral is characterized by a Latin cross plant with three naves, a false transept and a cancel (i.e., the area of the liturgical altar) in elevated position, due to the presence of a crypt containing the relics of Saint Geminianus.

It is of relevance to mention that, before the present Cathedral, three previous ones were built on the necropolis containing the tomb of St. Geminianus, which is the only remaining evidence of the first Cathedral. The second one was erected in the same place and the archeological remains indicate that this church had a length of around 32 m and width of 18 m. Finally, the presence of polylobate pillars, discovered during past excavations, allow to suppose the presence of another Cathedral, presumably built around the 11th century (Labate, 2009; Silvestri, 2013). As a consequence of the sequences of construction and demolition of previous Cathedrals, since the soil has "memory" of its previous loading history (Lancellotta, 2009, 2013), the Lanfranco and Wiligelmo Cathedral suffered uneven settlements not only moving from south towards north (as expected due to the interaction with the Tower), but also moving from east towards west.

Next to the Cathedral is the Ghirlandina Tower, the construction of which proceeded in parallel with that of the Cathedral. The tower is 88.82 m high and its squared hollow cross section is thicker on the corners because of the presence of four masonry pillars (Figure 2). In the inner part, an open stair runs along the walls and the belfry and the spire roof complete the structure. There are three horizontal masonry diaphragms built in the tower: the vault on the first floor, the floor of the Torresani cell and the vault above the belfry (the top deck instead is a timber structure). At ground level, two masonry arches connect the tower with the adjacent Cathedral (Figure 3).

The construction of the tower started at the same time as the cathedral in 1099 (see Dieghi, 2009; Piccinini, 2009). Thanks to delays and interruptions that allowed the foundation soil to increase its strength over time, the first five floors were successfully standing in 1169. The tower was finished in 1319, and the arches connecting the southern side of the tower to the cathedral, probably to prevent additional tilting of the tower towards the cathedral, were already in place in 1338. There is in fact evidence that the tower began to tilt during the construction, because the masons corrected the verticality of the walls accordingly.

Investigations aimed at clarifying the founding depth and the presumed presence of piles, as well as measurements of tilt, started at the end of the 19th century. At that time (from 1898 to 1901) a pit was initially excavated near the southern side of the tower and the inspection revealed the *socle* of the ancient door at depth of 1.80 m from the ground surface. Then, a trench was safely excavated at the North-East edge up to the depth of 4.90 m, where the *basolato* of the Roman road (via Emilia) was found. By direct inspection the presence of the socle was observed at a depth of 1.36 m and the masonry of the foundation was observed to reach a depth of 5.45 m. The conclusion was reached about the absence of any piled foundation, but this was rather arbitrary because of lack of any investigation beneath the founding level.

The Scientific Committee held in 2007 realized the need to have a deeper knowledge of what the tower foundation is like, so that we performed the borings sketched in Figure 4. This investigation allowed to reach the following conclusions: (a) the brickwork made foundation has a thickness of 3 m and was conceived as a spread foundation without supporting piles; (b) the socle of the ancient door was found at a depth of 1.48 m from the actual ground level; (c) the boring G5 was intentionally drilled in such a way to intercept the *basolato* of the Roman road (via Emilia) near the edge (at depth of 5.45 m) and just below the foundation (at depth of 6.75 m).

By comparing the difference (6.75–1.48) with that (4.90–1.36) found during the investigation made in 1898–1901, we can argue that the tower settlement was of the order of 1.73 m at the North side and, by considering the tilt, the average settlement was of the order of 1.85 m. This value represents a lower bound of the settlement the tower experienced, because we can certainly speculate that the depth of the *basolato* near the side of the foundation cannot be considered as representative of a free-field condition.

Similar arguments, linked to the depth of the *basolato* when referred to free field conditions would suggest an upper bound value of the average settlement equal to 2.07 m.

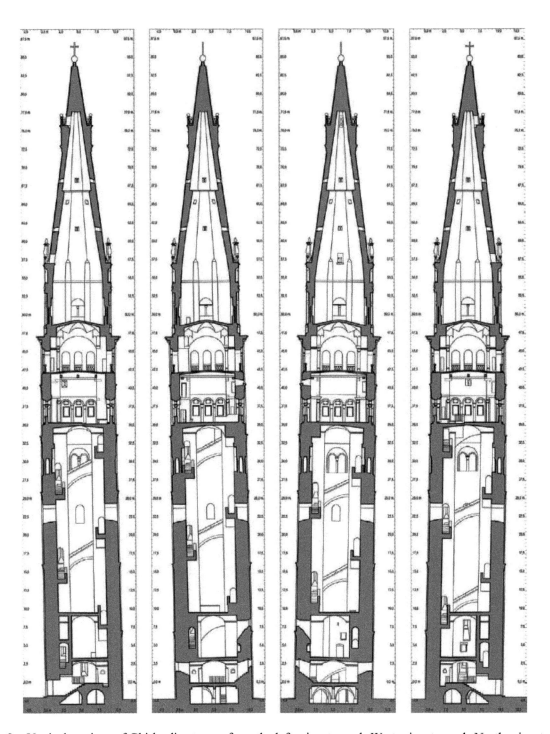

Figure 2. Vertical sections of Ghirlandina tower: from the left, view towards West, view towards North, view towards South, view towards East (Giandebiaggi *et al.*, 2009).

3 GEOLOGICAL SETTING, SOIL PROFILE AND GEOTECHNICAL CHARACTERIZATION

The Modena area is located in the southern margin of the Po plain, which represent the syntectonic sedimentary infill of the Plio-Pleistocene Apennine foredeep. The evolution of the Po basin is characterized by the transition from Pliocene open marine deposits to Quaternary marginal marine sediments followed by alluvial deposits. The Quaternary alluvial sequence shows a rhythmical alternation of coarse-grained (gravel and sand) and fine grained (silt and clay) deposits that are related to glacial-interglacial cycles with an overall cyclicity of 100.000 years (Lugli *et al.*, 2004).

Figure 3. The gothic arches connecting the Ghirlandina tower and the Cathedral (first constructed in 1338 and rebuilt in 1905).

The Pliocene–Quaternary succession of the Po Basin has been divided into six unconformity-bounded stratigraphic units (UBSU) marking phases of basin reorganization as a consequence of tectonic activity. The uppermost UBSU in the Po Basin is the "Emilia-Romagna Supersynthem" (middle-late Pleistocene and Holocene) in turn divided into a Lower and an Upper Emiliano-Romagnolo Synthem. The Upper Emiliano-Romagnolo Synthem is divided into sub-synthems, consisting of depositional cycles of silt-clay interlayered with gravel-sand (Cremaschi and Gasperi, 1989).

The uppermost 100 m of sediments in the Modena area belong to (from the bottom to the top):

1. the Bazzano Subsynthem (end of the middle Pleistocene up to about 120 ka),
2. the Villa Verucchio Subsynthem, which consists of the interglacial fine-grained sediments of the Niviano unit (about 120.000 to >30.000 years) and the last glacial coarse-grained deposits of the Vignola unit (>30.000 to >15.000 years)
3. the Ravenna Subsynthem, consisting of post-glacial fine-grained sediments (>15.000 years to present); its uppermost portion is called Modena unit (post-5th century A.D. to present).

Since 1980 the City of Modena promoted studies and investigations related to the subsidence of the Modena alluvial plain (Russo, 1985; Cancelli, 1986; Cancelli and Pellegrini, 1984; Pellegrini and Zavatti, 1980). Significant changes of piezometric levels within the major aquifers occurred during the mid 1970s due to municipal and industrial water supply purposes, lowering down to 10 m the original level and triggering a subsidence phenomenon of up to 80 to 90 cm in the town centre. This claimed for controls in order to reduce or prevent further negative effects, so that the piezometric level of the two aquifers, ranging from 22 to 34 m and from 54 to 63 m (see Figure 5), rose progressively at the end of the 1980s to a depth of 4 m from ground level and is at present about 0.5 meter above the ground level. Within the first horizon the groundwater level oscillates from a depth of 1.50 to 1.2 m.

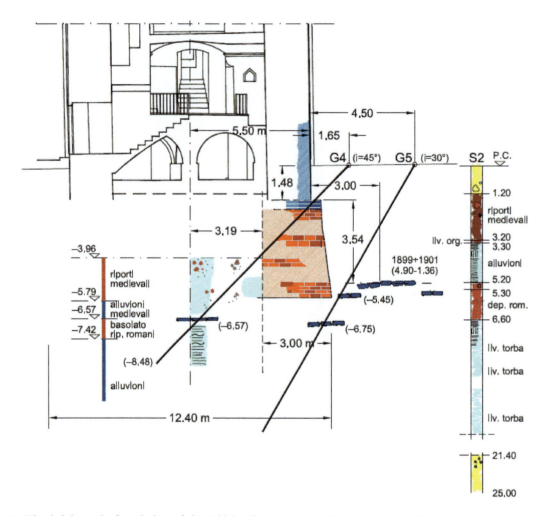

Figure 4. The brick-made foundation of the Ghirlandina tower and the "basolato" of the Roman road "via Emilia" (Lancellotta, 2009).

In addition to these studies, it is relevant to outline that the Modena alluvial plain is characterized by a unique abundance of archaeological sites, and the related interest promoted researches on the Quaternary sedimentation of the Modena plain (Cremaschi and Gasperi, 1989; Fazzini and Gasperi, 1996; Lugli et al., 2004).

Finally, in order to define a detailed soil profile and the relevant mechanical parameters aimed at investigating the behaviour of the Ghirlandina Tower in relation to subsoil conditions, a rather comprehensive site investigation was planned in September 2007 and December 2008.

By referring to Figure 5, the soil profile down to the investigated depth of 80 m is composed of a first horizon of medium to high plasticity inorganic clays 22 m-thick, with abundance of laminae of sands and peat, only a few millimetres-thick. The upper portion of this horizon, whose thickness ranges from 5 to 7 meters, belongs to the *Modena unit* and is linked to the flooding events (of post-Roman age) mostly produced by minor streams (the Fossa-Cerca stream).

The underlaying horizons, ranging from depth of 22 to 54 meters, represent the result of a complete transgressive-regressive cycle: the fine-grained sediments belong to the *Niviano unit*, and were deposited during the penultimate interglacial cycle, and the overlying coarse-grained materials, the *Vignola unit*, were mostly deposited by the Secchia river during the last glacial peak.

A second horizon of coarse-grained materials is encountered at depths ranging from 54 to 63 m, and thereafter a fine-grained horizon is found down to the depth of 78 m, here again characterized by a diffuse presence of sand laminae. This coarse-/fine-grained sequence belongs to the Bazzano Subsynthem (up to 120.000 years ago) and its coarse-grained upper part marks the penultimate glacial event.

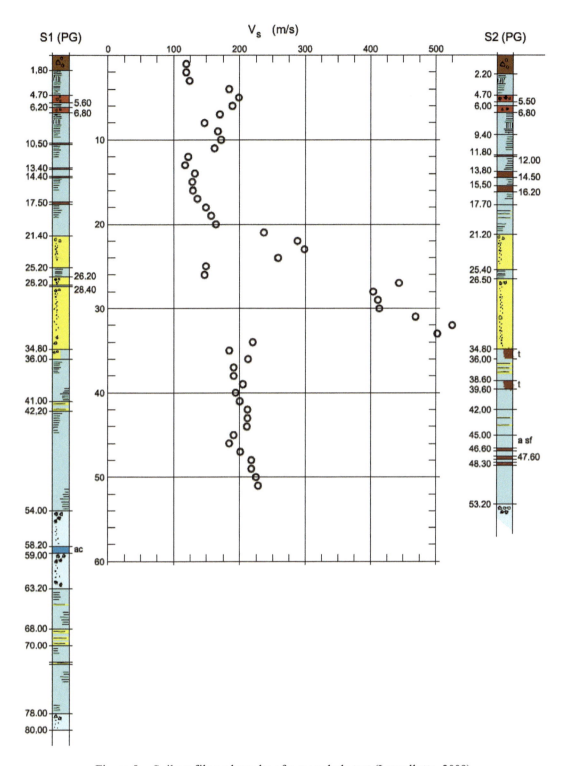

Figure 5. Soil profile and results of a cross-hole test (Lancellotta, 2009).

The geotechnical activity has been finally complemented with the execution of cross-hole tests (Figure 5). In this respect, we remind that shallow seismic exploration tests of soils represent an important class of field tests, because of their non invasive character. This allows to preserve the initial structure of soil deposits as well as the influence of all diagenetic phenomena contributing to a stiffer mechanical response. Soils are tested at small strain level, i.e. less than 10^{-3}%, and a reasonable assumption is to analyze the results of these tests by referring to the wave propagation theory in elastic materials. Therefore, the cross-hole test represents one of the most reliable methods of determining the shear modulus at small strain amplitude.

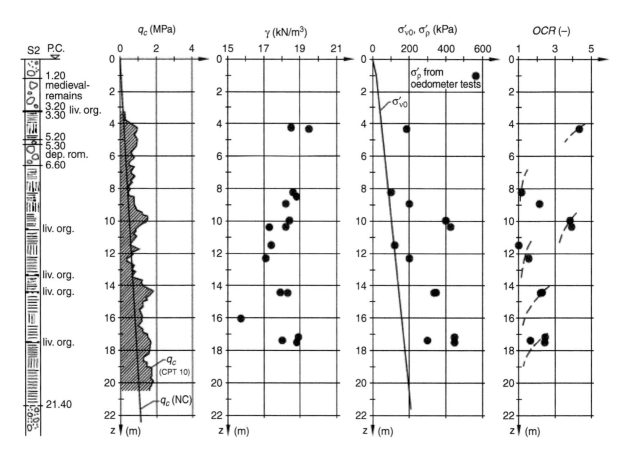

Figure 6. Stress history of Modena clay (Lancellotta, 2009) (γ = unit weight; q_c = cone resistance).

It is well known that the assessment of the over-consolidation stress σ'_p is the most important step when characterizing the soil mechanical behaviour and it is of relevance to recall that the trend of yield stress and overburden effective stress σ'_{vo} with depth represents a concise picture of the stress history of the deposit (Jamiolkowski et al., 1985).

Different mechanism can obviously act together and quite often the stress history profiles are not so simple. This is the case under consideration, because when considering the trend of yield stress and overburden effective stress with depth shown in Figure 6, as well as the trend of the over-consolidation ratio $OCR = \sigma'_p/\sigma'_{vo}$ on the same figure, no simple arguments are able to justify this apparently erratic trend. More insight in this case was gained by considering the trend of cone penetration resistance (q_c). Figure 6 reports a representative example of these tests and on the same figure we superimpose the expected trend in a virtually normally consolidated soil deposit. Since it is well known that in a normally consolidated soil the cone resistance linearly increases with depth, any deviation from this trend is certainly matter of over-consolidation phenomena. In addition, by observing that such deviations have a cyclic character, it can be argued that the presence of over-consolidated soil layers can be related to temporary exposure and desiccation during the deposition cycles.

In order to investigate the influence of soil non linearity in soil-structure analyses, the cyclic behavior of the material at larger strains has been assessed through resonant column tests performed on high quality undisturbed samples, and Figure 7 summarizes the obtained results in terms of decay of shear modulus G (normalized with respect to the initial value G_o) with shear strain γ, as it is customary in Soil Mechanics to represent soil non linearity.

4 SOIL-STRUCTURE INTERACTION AND IDENTIFICATION ANALYSES

The advent of powerful digital computers and the development of advanced numerical methods seems today allow to analyze any difficult interaction problems. However, the use of the so called "direct approach", which

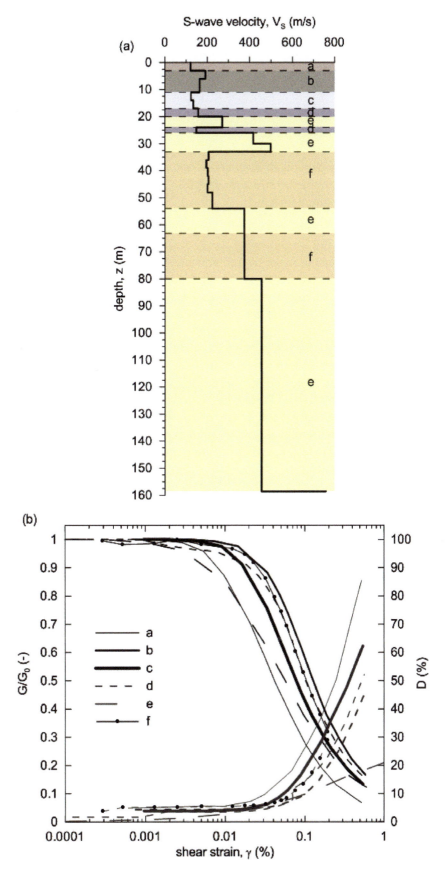

Figure 7. Results of resonant column tests on undisturbed samples (small letters indicate the depth of sampling, shown in the upper part) (Lancellotta, 2009; Cosentini *et al.*, 2015).

simulates the complete dynamic soil-structure interaction, requires a lot of expertise that is still far beyond the current engineering practice.

These difficulties motivated the introduction and the use of the so called "lumped parameters method" (Sarrazin *et al.*, 1972), rooted on the complete rigorous solution provided by Veletos and Wei (1971) and Luco and Westmann (1971). This procedure considers a "decoupled problem" to be solved in three steps (Gazetas, 1991):

1. the *kinematic interaction*, that analyses the response of the foundation to the actual seismic motion defined in the free-field;
2. the formulation of the frequency dependent *dynamic impedances* for the foundation;
3. the *inertial interaction*, that analyses the structure supported on the impedances, defined in step 2, and subjected to the base motion obtained in step 1.

As far as the definition of dynamic impedances, parametric studies made by Sarrazin *et al.* (1972), Roesset *et al.* (1973) and Veletsos and Meek (1974) allowed to clarify the relative importance of the swaying (horizontal translation) and rocking functions and to make the remarks here briefly recalled.

For usual high rising structures, the frequency associated to rocking oscillations is smaller than the swaying frequency, and this latter is normally higher than the fundamental frequency of the structure.

The percentage of critical radiation damping in rocking is usually very small (few percent); on the contrary that of the critical damping in swaying is very large (up to 40% or more). As a consequence, for slender structures, as it is the case of Ghirlandina tower, the effect of swaying spring is practically negligible; on the other end, for rigid structures, when a small value of swaying damping ratio is specified, the results show unrealistically large amplifications and it could be more realistic to neglect the horizontal spring.

As suggested by Veletsos and Meek (1974), in order to assess the influence of the interaction with the supporting soil, characterized by a density ρ and a shear wave velocity V_S, the structure can be represented as a single degree of freedom oscillator, with mass M, height h and flexural stiffness K properly defined. If the fixed-base period of the structure is $T_o = 1/f_o$ and its damping ratio is ξ, the effect of soil-structure interaction is to modify these parameters as follows

$$T^* = T_o\sqrt{1 + \frac{K}{K_x} + \frac{Kh^2}{K_\alpha}} \tag{1}$$

$$\xi^* = \left(\frac{T_o}{T^*}\right)^3 \left[\xi + \frac{(2-v)\,\pi^4\delta}{2\sigma^3}\left(\frac{\beta_x}{\alpha_x^2}\frac{R^2}{h^2} + \frac{\beta_\vartheta}{\alpha_\vartheta^2}\right)\right] \tag{2}$$

where

$$\sigma = \frac{V_S}{hf_o} \tag{3}$$

$$\delta = \frac{M}{\rho\pi R^2 h} \tag{4}$$

and R is the radius of the equivalent circular foundation. The coefficients α_j, β_j define the impedance functions (Gazetas, 1991)

$$\tilde{K}_j = k_j + i\omega\,c_j = K_j(\alpha_j + i\beta_j a_o) \tag{5}$$

K_j being the static stiffness (the index j stands for swaying or rocking) and a_o is introduced to render dimensionless the frequency ω of the harmonically excitation

$$a_o = \frac{\omega R}{V_S} \tag{6}$$

It appears from equation (1) and (2) that the equivalent period increases with respect to the fixed base value, independently on the slenderness of the towers, whereas the damping ratio may decrease or increase respectively for slender or squat towers.

In addition to these remarks, it must be stressed that, due to soil highly non linear behavior, the shear modulus G, used to evaluate the soil foundation stiffness, should be based on a representative shear strain level. No practical guidelines can be found in literature concerning this aspect, but for earthquake loading it appears rather realistic a value of G consistent with the shear strain level computed from a seismic site response analysis.

Furthermore, even if the value of the rocking impedance can be obtained by using theoretical or numerical procedure, it is rather difficult to account for the 3D nature of the problem, as well as the effectiveness of the soil contact along the vertical sides of an embedded foundation.

In this respect, a rather significant contribution comes from experimental identification analyses, that allow to estimate the dynamic parameters of the structure such as the natural frequencies of vibration, mode shapes, and damping (Ljung, 1999).

These dynamic tests are based on measuring, analyzing and processing vibrations induced by inertial forces and can be performed using known or unknown inputs.

In the present case, dynamic tests were performed with ambient excitation, to preserve the integrity of the structure and to allow a continuous long term monitoring (Lancellotta and Sabia, 2014). In particular, to properly capture the higher modes of vibration, triaxial accelerometers were installed on the tower to record the time histories along three directions as shown in Figure 8 (X: normal to the cathedral nave; Y: vertical; Z: parallel to the cathedral nave).

To select the optimal position of the sensors, a preliminary dynamic analysis was performed using a finite element model of the tower composed of 24,296 solid elements. As expected, the results are sensitive to many parameters, the most relevant of which are the soil-foundation stiffness, the mechanical properties of the masonry, the abrupt change of geometry of the tower section and the constraints offered by the arches connecting the tower and the cathedral. Therefore an analysis was performed to detect the sensitivity of mode shapes to these parameters, in order to avoid ambiguous estimates resulting from insufficient points or incorrect distribution of these points on the structure (spatial aliasing).

The modal identification was carried out in two steps: first, by analyzing the signals detected in the horizontal directions, X and Z; second, by analyzing the vertical component Y. The selection of the frequencies of vibration and the corresponding mode shapes was made on the basis of the recurrent forms, with an equivalent viscous damping less than 10% and a level of affinity higher than 90%, evaluated by means of the "Modal Assurance Criterion" (MAC) .

The analysis allowed the estimation of the natural frequencies, mode shapes, and damping parameters. Examples of mode shapes are shown in Figure 9, with reference to both bending (modes 1 to 7) and extensional modes, these latter being vibrations along the tower axis, (modes 8 and 9).

These results prove that the soil–structure interaction cannot be neglected in contrast to most published identification analyses, which usually assume the structure to have rigid constraints at its base. Indeed, the first bending mode shape (Figure 9.a) associated to a frequency of 0.74 Hz shows a rotation and displacement pattern at the tower base due to soil deformability. On the contrary, the second bending mode shape (Figure 9.b), associated to a frequency of 0.85 Hz, reflects the presence of the arches connecting the tower and cathedral. The influence of soil-structure interaction is also shown by the axial modes of vibration. For the first axial mode (mode 8 in Figure 9.c), the base moves in phase with the top of the tower, and about 60% of the associated vertical modal displacement is determined by soil deformability. The second axial mode (mode 9) shows displacements of the base in opposite phase with respect to the top of the tower. Based on these results, Lancellotta and Sabia (2013) proved that the response of the soil-structure system can be optimized by using a masonry modulus Em ranging from 3 to 4 GPa and the following values of the rocking stiffness $K_\alpha = 240\,\mathrm{GN} \cdot \mathrm{m}$.

It is of interest to compare this value with the expected one based on classical approaches of vibrations of foundations (Gazetas, 1991). Considering the width of the basement and the cross-hole results reported in Figure 5, a representative value of the shear wave velocity for the foundation soils is $V_S = 125\,\mathrm{m/s}$, from which the small-strain shear modulus is estimated as:

$$G_o = \rho v_s^2 = 28\,\mathrm{MPa}$$

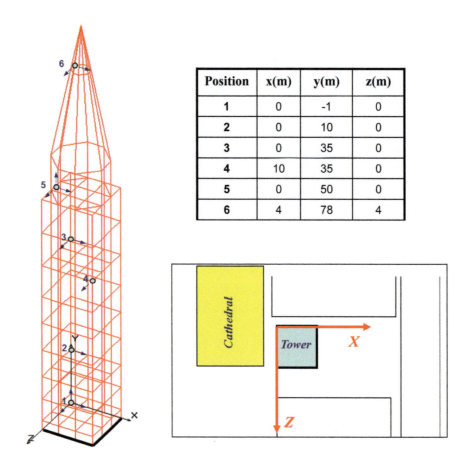

Figure 8. Layout of measurement points (Lancellotta and Sabia, 2013).

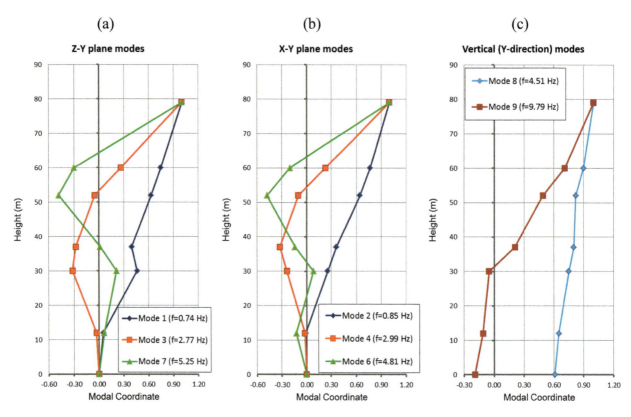

Figure 9. Bending and axial modal shapes (Lancellotta and Sabia, 2013).

This value refers to free field conditions, so that it has been corrected in order to account for the stress level induced by the tower, therefore the representative values being $G_o = 44$ MPa. The rocking stiffness is then estimated as (Gazetas, 1991):

$$K_\alpha = \frac{3.6 G b^3}{1 - \upsilon} \cdot f_D = 240 \, \text{GN} \cdot \text{m}$$

where f_D is the correction for embedment of the foundation, given by:

$$f_D = \left\{ 1 + 1.26 \frac{d}{b} \left[1 + \frac{d}{b} \left(\frac{D}{d} \right)^{0.2} \sqrt{\frac{b}{l}} \right] \right\} = 3.19$$

in which:

$2b$ and $2l$ are the dimensions of the foundation ($2b = 2l = 12.40$ m);

D is the founding depth;

d is the fraction of D that contributes to the constraint (here assumed equal to D).

The circumstance that the obtained value is consistent with that provided by the identification analysis, represents a sound validation of the analytical approach.

By considering the very low strain level involved in both the identification analysis and the shear wave propagation in cross-hole tests, this value is deemed to be appropriate only for a low intensity seismic motion. Further considerations are therefore needed when dealing with a strong motion in order to account for the non-linearity of soil behaviour, as it is further discussed.

5 SOIL-STRUCTURE INTERACTION TAKING INTO ACCOUNT SOIL NON-LINEARITY

To study the role of soil–structure interaction in the long-term performance and to assess the seismic vulnerability of the tower, a dynamic monitoring program was started in August 2012. The acquisition system operates with a sampling frequency of 100 Hz and allows continuous monitoring of the dynamic response of the tower under ambient vibrations. In the following reference is made to time histories of acceleration during three seismic events (Figure 10): October 3rd, 2012 (epicenter in Piacenza and magnitude M = 4.5), January 25th, 2013 (epicenter in Garfagnana, M = 4.8) and June 21st, 2013 (epicenter in Alpi Apuane, near Lucca, M = 5.2).

The same events have also been recorded in free field conditions at the Modena station of the Italian Accelerometric Networks, as shown in Figure 11. The shear wave velocity profile for the site of the station MDN, was estimated on the basis of passive surface wave tests (see Foti *et al.*, 2011 for further details) and reported in Figure 12.

The free-field records have then been deconvoluted to obtain the reference ground motions for bedrock, the latter assumed at a depth of about 160 m, where the shear wave velocity reaches the value of 800 m/s. Considering the relatively small distance between MDN station and the Ghirlandina Tower, if compared to the epicentral distances, this reference motions was used to study the seismic response of the Ghirlandina site without the need to account for any attenuation effect.

The seismic ground response analysis was performed by using an *equivalent linear approach* (Idriss and Seed, 1968), as implemented in the numerical code EERA (Bardet *et al.*, 2000). The reference shear wave velocity profile (Figure 7) has been estimated on the basis of the cross-hole test of Figure 5, complemented with the information reported in Figure 7, as far as soil non-linearity is concerned. As already mentioned, the reduction of normalized shear modulus and increment of damping ratio with strains have been measured on undisturbed samples with resonant column tests for the clayey layers (Figure 7). For the sandy layers (letter "e" in Figure 7), reference was made to the shear modulus reduction curves suggested by Seed and Idriss (1970). Also note that the sample obtained for the most shallow layer (letter "a" in Figure 7) shows a lower linear threshold than other samples in agreement with its different grain-size distribution.

The results of the seismic ground response analysis for the Ghirlandina tower site are reported in Figure 13 for the three events recorded by the accelerometric station MDN. The profiles of maximum shear strains, if

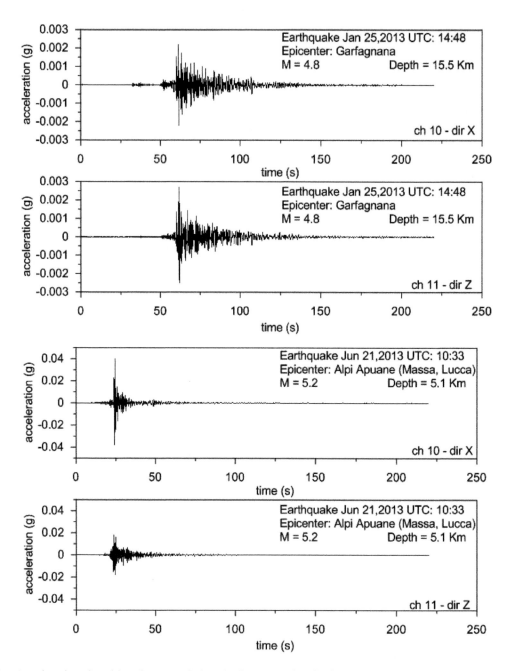

Figure 10. Acceleration time histories recorded at the basement level of the tower during the earthquake of January 2013 and June 2013 (horizontal components) (Cosentini *et al.*, 2015).

compared with experimental reduction curves of the normalized shear modulus (G/G_0) (Figure 7), show that for the first two events the response is associated to the behaviour below the linear threshold strain. The third event instead caused larger strains, beyond the linear threshold, and in particular the largest deformations are attained in the zone below the foundation, where a weak soil horizon is encountered. This layer is likely to have a major impact on the rocking response of the tower.

These results suggest that an estimate of the operational shear modulus could be obtained directly from the seismic site response analysis, and in the present case, consistently with the shear strain level caused by the seismic motion, an operative value of G/G_0 ranging from 0.75 to 0.8 is obtained within the zone of influence beneath the foundation.

To validate the suggested procedure, an independent estimate of the reduction of foundation stiffness with increasing seismic action has been made by means of Frequency Response Functions (FRF) defined as follows

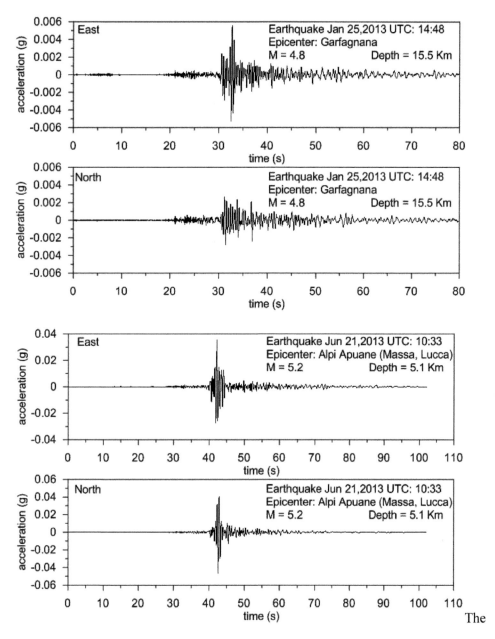

Figure 11. Acceleration time histories for the seismic events of January 2013 and June 2013 as recorded by the MDN station of the Italian accelerometric network (downloaded from http://itaca.mi.ingv.it) (Cosentini *et al.*, 2015).

(Bendat and Piersol, 1993; Ewins 2000).

$$H(\omega) = \frac{Y(\omega)}{X(\omega)} \qquad H(\omega) = \frac{S_{XY}(\omega)}{S_{YY}(\omega)}$$

where $Y(\omega)$ and $X(\omega)$ are the Fourier transforms of the output and of the input, respectively; $S_{XY}(\omega)$ is the cross-spectra of the input $x(t)$ and the output $y(t)$; $S_{YY}(\omega)$ is the auto-spectra of the input $y(t)$.

The FRF is a complex function which amplitude has a maximum at the resonant frequency, where the output is 90° out of phase with respect to the input. The magnitude of the modal coefficient is simply taken as the value of the imaginary part at resonance. The sign is either positive or negative, considering the direction of the peak along imaginary axis. This implies that the phase angle is either 0° or 180°.

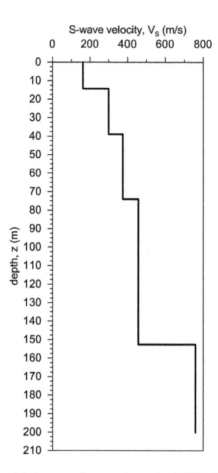

Figure 12. V_S profile at the site of the Modena accelerometric station MDN (http://itaca.mi.ingv.it) (Cosentini et al., 2015).

Figure 14 compares the FRF for the three seismic events (October 2012, January 2013 and June 2013), and shows a reduction of the first natural frequency of the tower (from 0.74 to 0.69 Hz) moving from the first two events to that of June 2013.

This difference is certainly associated to soil non-linearity, since significant structural non-linearity is not expected for the masonry walls for such a small seismic excitation. Therefore, the observed difference in natural frequency can then be converted into an estimate of the reduction of the foundation stiffness by using the following arguments.

Let assume the tower to be represented by an equivalent single degree of freedom model, with a mass lumped at an height h over the base of the foundation and a structural stiffness equal to K_s. If T_o is the fundamental period of the structure on a rigid base, it can be proven that the period of the structure-soil system increases when the flexibility of the soil is taken into account and is given by (the swaying term is negligible in equation 1)

$$T = T_o\sqrt{1 + \frac{K_s h^2}{K_\alpha}}$$

where K_α is the rocking stiffness of the soil-foundation system.

Provided that the value of T_o can be obtained by a numerical model calibrated on the structural identification process (in the present case $T_o = 1.01$ s), the ratio between the mobilized soil stiffness during two different seismic events can be obtained by using the inverse formula

$$\frac{K_{\alpha 2}}{K_{\alpha 1}} = \frac{(T_1/T_o)^2 - 1}{(T_2/T_o)^2 - 1}$$

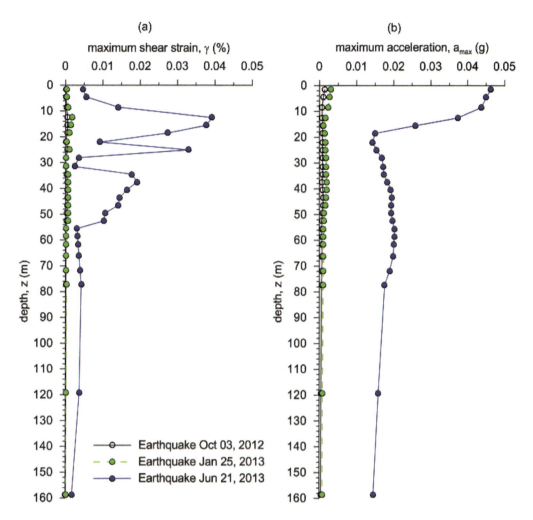

Figure 13. Seismic ground response as estimated with the equivalent linear analysis for the seismic events recorded by MDN accelerometric station: a) profile of maximum shear strains; b) profile of maximum acceleration (Cosentini *et al.*, 2015).

From the difference in fundamental frequency observed in Figure 14b (here reference is made to Z-direction because it is not influenced by the arches connecting the Tower to the Cathedral), a stiffness ratio equal to 0.78 is obtained, consistently with the shear strain level derived from the seismic ground response analysis (Figure 13).

6 REMARKS

(a) The preservation of cultural heritage requires a careful assessment of the necessary remedial measures. It is indeed essential to preserve not only the shape and appearance of a monument but also its historical and material integrity. In this context, the construction techniques, the materials, and the structural scheme need to be considered within a multidisciplinary approach. A patient attitude is necessary to devise the stabilizing measures only after the behavior of the monument is properly understood.

In particular, assessing seismic vulnerability and long-term behavior of historical towers requires modeling the interaction of the structure with the supporting soil, because soil-structure interaction strongly affects the seismic capacity of the system, related to soil strength, as well as the seismic demand (Di Tommaso *et al.*, 2013).

In the latter case, depending on the soil stiffness, the fundamental period of the structure will increase, with a corresponding reduced spectral acceleration. This reduction, combined with energy dissipation

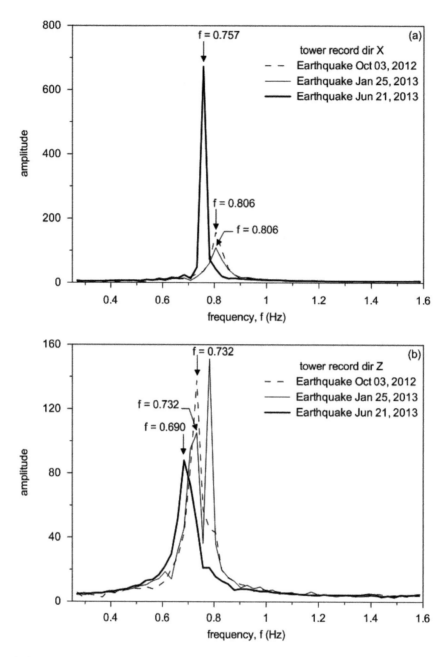

Figure 14. Cumulative Frequency Response Function of the tower obtained as the ratio between records at the different elevations and the basement (Cosentini et al., 2015).

mechanisms, give reason for the good performance of the Ghirlandina tower during past seismic events (1501, Castelvetro; 1505, Bologna; 1671, Rubiera; 1832, Reggio Emilia; 1996, Correggio), despite the damage suffered by the Cathedral.

(b) Experimental identification analyses performed under ambient vibration excitation provide a sound validation of theoretical approaches suggested in literature to estimate the dynamic stiffness of a soil-foundation system. However, considering the very low strain level involved in both identification analysis and shear wave propagation during cross-hole tests, the obtained values of the soil-foundation stiffness is appropriate only for low intensity seismic motions. Further considerations need to be introduced when dealing with strong motions in order to account for the non-linear soil behavior. Therefore, a value of G consistent with the shear strain level computed from a seismic site response analysis should be used.

(c) The above conclusion applies to the present case, since it has been validated by means of a continuous monitoring system. This aspect deserves special attention because, as it was shown, monitoring increases the capability to detect the importance of non-linear phenomena as well as possible damage that could occur in the long term performance of the structure, subjected to repeated seismic events.

REFERENCES

Bardet, J. P., Ichii, K. and Lin, C. H. (2000). A Computer Program for Equivalent-linear Earthquake site Response Analyses of Layered Soil Deposit. User's manual, University of Southern California.

Bendat, J.S. and Piersol, A.G. (1993). Engineering applications of correlation and spectral analysis, New York. Wiley-Interscience.

Binda, L., Gatti, G., Mangano, G., Poggi C. and Sacchi Mandriani, G. (1992). The collapse of the Civic Tower of Pavia: a survey of the materials and structure. *Masonry International*, **20**, (6), 11–20.

Brandi, C. (1977). Teoria del restauro, Einaudi, 154 pp. 1st edn, 1963, in Edizioni di Storia e Letteratura, Torino, Einaudi.

Brandi, C. (1985). Disegno dell'architettura italiana, 293 pp. Torino, Einaudi.

Burland, J.B., Jamiolkowski, M. and Viggiani, C. (2003). The stabilization of the leaning Tower of Pisa, *Soils and Foundations*, **43**, (5), 63–80.

Calabresi, G. (2013). The role of Geotechnical Engineers in saving monuments and historic sites. Kerisel Lecture, in Proceeding of XVIII International Conference on Soil Mechanics and Geotechnical Engineering (ed. P. Delage, J. Desrues, R Frank, A. Puech, F. Schlosser), 71–83, Paris.

Calabresi, G. and D'Agostino, S. (1997). Monuments and historic sites: intervention techniques, in Proc. of Int. Symp. on Geotechnical Engineering for the preservation of Monuments and Historic Sites, (ed C. Viggiani, Balkema), 409–425.

Cancelli, A. (1986). Aspetti geotecnici della subsidenza. Ambiente: Protezione e Risanamento. Atti 2° Corso di Aggiornamento per tecnici di igiene ambientale. Modena. Ed. Pitagora, Bologna.

Cancelli, A. and Pellegrini, M. (1984). Problemi geologici e geotecnici connessi al territorio della città di Modena. Atti 2° Congr. Naz. ASS.I.R.C.CO., Ferrara, 53–64.

Como, M. (1993). Plastic and visco-plastic stability of leaning towers, Int. Conf. Physic-Mathematic and Structural Engineering, in memory of G. Krall, Elba.

Cosentini, R.M., Foti, S., Lancellotta, R. and Sabia, D. (2015). Dynamic behaviour of shallow founded historic towers: validation of simplified approaches for seismic analyses. International Journal of Geotechnical Engineering, V.9, 1, 13, W.S. Maney et Son Ltd.

Cremaschi, M. and Gasperi, G. (1989). L'alluvione alto medioevale di Mutina (Modena) in rapporto alle variazioni ambientali oloceniche. Mem. Soc. Geol. It., 42, 179–190.

Desideri, A. and Viggiani, C. (1994). Some remarks on the stability of towers. Symp. on Development in Geot. Eng. (from Harvard to New Delhi, 1936–1994), Bangkok.

De Silva, F., Ceroni, F., Sica, S., Pecce, M.R. and Silvestri, F. (2015). Effects of soil-foundation-structure interaction on the seismic behavior of monumental towers: the case study of the Carmine Bell Tower in Naples. Rivista Italiana di Geotecnica, XLIX, 3, 7–27.

Dieghi, C. (2009). Fonti e studi per la storia della ghirlandina, in Cadignani R. (ed.) La Torre Ghirlandina: un progetto per la conservazione, L. Sossella Publisher, vol. 1, 48–65, Roma.

Di Tommaso, A., Lancellotta, R., Sabia, D., Costanzo, D., Focacci, F. and Romaro, F. (2013). Dynamic identification and seismic behaviour of the Ghirlandina Tower in Modena (Italy). Proc. 2nd Int. Symposium on Geotechnical Engineering for the Preservation of Monuments and Historic Sites, Taylor & Francis, 343–351.

Ewins, D. J. (2000). Modal Testing – theory, practice and application, Research Studies Press LTD. Baldock, Hertfordshire, England

Fazzini, P. and Gasperi, G. (1996). Il sottosuolo della città di Modena. Accad. Naz. Sci. Lett. Arti di Modena, Miscellanea Geologica, 15, 41–54.

Federico, F. and Ferlisi, S. (1999). Time evolution of stability of leaning towers. VI Int. Conf. STREMAH, Dresden, Brebbia and Jager Edrs, 485–494, WIT Press.

Foti, S., Parolai, S., Bergamo, P., Di Giulio, G., Maraschini, M., Milana, G., Picozzi, M. and Puglia, R. (2011). Surface wave surveys for seismic site characterization of accelerometric stations in ITACA. *Bulletin of Earthquake Engineering, Springer*, **9**, 1797–1820.

Gazetas, G. (1991). Foundation vibrations. In H.F. Fang, Foundation Engineering Handbook, (ed. Van Nostrand Reinhold), 553–593, New York.

Giandebiaggi, P., Zerbi, A. and Capra, A. (2009). Il rilevamento della torre Ghirlandina, in La Torre Ghirlandina: un progetto per la conservazione, (ed. L. Sossella), vol. 1, 78–87, Roma.

Hambly, E.C. (1985). Soil buckling and the leaning instability of tall structures, *The Structural Engineer*, **63**, 3, 77–85.

Heyman, J. (1992). Leaning towers, *Meccanica*, **27**, 153–159.

Idriss, I. M. and Seed, H. B. (1968). Seismic Response of Horizontal Soil Layers, *Journal ofthe Soil Mechanics and Foundations Division, ASCE*, **94**, (4), 1003–1031.

Jamiolkowski, M., Ladd, C.C., Germaine, J.T. and Lancellotta, R. (1985). New developments in field and laboratory testing of soils. Theme Lecture XI ICSMFE, San Francisco, 1, 57–152.

Labate, D. (2009). Archeology's contribution to understanding a monument. In: Cadignani R. (ed.) *The Ghirlandina Tower – Conservation Project*. L. Sossella Publisher, Roma, 66–77. ISBN: 9788889829721.

Lancellotta, R. (1993). The stability of a rigid column with non linear restraint, *Géotechnique*, **33**, (2), 331–332.

Lancellotta, R. (2009). Aspetti geotecnici nella salvaguardia della torre Ghirlandina, in Cadignani R. (Ed.) La Torre Ghirlandina: un progetto per la conservazione, L. Sassella Publisher, vol. 1, 178–193, Roma.

Lancellotta, R. (2013). La torre Ghirlandina: una storia di interazione struttura-terreno. XI Croce Lecture, *Rivista Italiana di Geotecnica*, **2**, 7–37.

Lancellotta, R. and Sabia, D. (2013). The role of monitoring and identification techniques on the preservation of historic towers. Keynote Lecture, 2nd Int. Symposium on Geotechnical Engineering for the Preservation of Monuments and Historic Sites, Napoli, 57–74.

Lancellotta, R. and Sabia, D. (2014). Identification technique for soil-structure analysis of the Ghirlandina tower. *International Journal of Architectural Heritage*. DOI. 10.1080/15583058.2013.793438.

Ljung, L. (1999). System identification. Theory for users, N.J., Prentice Hall.

Luco, J.E. and Westman, R.A. (1971). Dynamic response of circular footings, *ASCE, Journal of Engineering Mechanics Division*, **97**, (EM5), 1381–1395.

Lugli, S., Marchetti Dori, S., Fontana, D. and Panini, F. (2004). Composizione dei sedimenti sabbiosi nelle perforazioni lungo il tracciato ferroviario ad alta velocità: indicazioni preliminari sull'evoluzione sedimentaria della media pianura modenese, *Il Quaternario, Italian Journal of Quaternary Sciences*, **17**, 379–389.

Macchi, G. (1993). Monitoring medieval structures in Pavia, *Structural Engineering International*, **1**, 6–9.

Marchi, M., Butterfield, R., Gottardi, G. and Lancellotta, R. (2011). Stability and strength analysis of leaning towers, *Géotechnique*, **61**, (12), 1069–1079.

Nova, R. and Montrasio, L. (1995). Un'analisi di stabilità del campanile di Pisa, *Rivista Italiana di Geotecnica*, **2**, 83–93.

Pellegrini, M. and Zavatti, A. (1980). Il sistema acquifero sotterraneo fra i fiumi Enza, Panaro e Po: alimentazione delle falde e scambi tra falde, correlazioni idrochimiche. Quad. Ist. Ric. Sulle Acque, CNR, 51, 205–217.

Piccinini, F. (2009). Note sul cantiere del Duomo e della Ghirlandina: Lanfranco, Wiligelmo, i Campionesi e il Comune medievale a Modena, in La Torre Ghirlandina: un progetto per la conservazione, (ed. L. Sossella), Vol. 1, 42–65, Roma.

Pisanò, F., di Prisco, C.G. and Lancellotta, R. (2014). Soil-foundation modelling in laterally loaded historical towers, *Géotechnique*, **64**, 1, 1–15.

Roesset, J.M., Whitman, R.V. and Dobry, R. (1973). Modal analysis of structures for foundation interaction, *ASCE, Journal of Structural Division*, **99**, (ST3), 399–416.

Russo, P. (1985). L'abbassamento del suolo nella zona di Modena (1950–1982). Tecnica Sanitaria, XXIII, 293-3

Sarrazin, M.A., Roesset, J.M. and Whitman, R.V. (1972). Dynamic soil-structure interaction, *ASCE, Journal of Structural Division*, **92**, (ST7), 1525–1544.

Seed, H.B. and Idriss, I.M. (1970). Soil moduli and damping factors for dynamic response analyses, Report EERC 70-10, Un. California, Berkeley.

Silvestri, E. (2013). Una rilettura delle fasi costruttive del Duomo di Modena. Atti e memorie, deputazione di storia patria per le antiche provincie modenesi 2013; s. XI, XXXV: 117–149.

Veletsos, A.S. and Meek, J.W. (1974). Dynamic behaviour of building-foundations systems. Earthquake Engineering and Structural Dynamics, III, 121–138.

Veletos, A.S. and Wei, Y.T. (1971). Lateral and rocking vibration of footings, *ASCE, Journal of Soil Mechanics and Foundation division*, **97**, 1227–1248.

Geotechnics and Heritage: Historic Towers – Lancellotta, Flora & Viggiani
© 2018 Taylor & Francis Group, London, ISBN 978-1-138-03272-9

The Big Ben Clock Tower: Protective compensation grouting operations

D.I. Harris
DI Harris Geotechnics Ltd., UK

R.J. Mair
University of Cambridge, UK

J.B. Burland & J.R. Standing
Imperial College London, UK

ABSTRACT: The construction of the Jubilee line Extension Station at Westminster, London was predicted to produce significant movements of the Big Ben Clock tower and the adjoining Palace of Westminster. The works consisted of the excavation of two vertically stacked 7.4 m tunnels and a 39 m deep station escalator box respectively located 28 m and 34 m north of the foundations of the Clock Tower. The protective measures adopted to minimise damage to these priceless historic buildings consisted primarily of compensation grouting below the Clock Tower and proved extremely effective in controlling settlement and tilt of the structure. This case study not only demonstrates the success of this protective measure, but also shows the value of careful interpretation of appropriate numerical modelling and the results of high quality monitoring during and after completion of the works.

1 INTRODUCTION

The Big Ben Clock Tower was constructed in 1858, and consists of load-bearing brickwork with stone cladding approximately 11 m square to a height of 61 m, supporting a cast iron framed belfry and spire to a total height of 92 m. The Tower is founded on a mass concrete raft, 15 m square and 3 m thick within Terrace Gravels overlying London Clay. The weight of the Tower is about 8400 t giving an average foundation bearing pressure of approximately 400 kPa. It is worth noting that Big Ben is the name of the largest bell in the belfry. The clock tower itself was known as St. Stephen's Tower but it was recently re-named the Elizabeth Tower.

The Clock Tower is structurally connected to the four-storey East Wing of the Palace of Westminster, as shown in Figure 1, which houses the offices of the Ministers of State. Both the Clock Tower and the Palace have a single-level basement of vaulted brickwork. Like the Clock Tower, the Palace of Westminster is founded on a mass concrete raft.

The construction of the new Westminster Station on London Underground Limited's Jubilee Line Extension (JLE) project was predicted to cause significant movements of the Clock Tower and the adjoining Palace of Westminster. The station consists of two station platform tunnels, one vertically above the other, and a 39 m deep station escalator 'box' for access purposes – see Figure 1. Protective measures, primarily in the form of compensation grouting beneath the Clock Tower, were implemented during the construction period to control the settlement and tilt of the monument. This chapter describes these protective measures and presents the results of the monitoring during and subsequent to the works. Figure 2 is a photograph of the Palace of Westminster taken during the construction of the station.

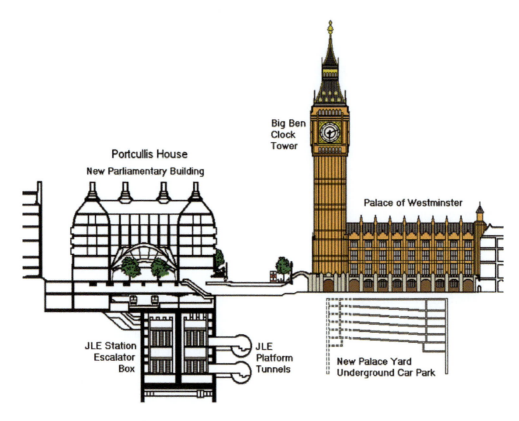

Figure 1. Cross-Section showing Westminster JLE Station, Big Ben Clock Tower the Palace of Westminster.

Figure 2. View of Palace of Westminster during construction of Westminster Station.

2 THE NEW WESTMINSTER STATION AND ITS CONSTRUCTION

The layout of the new Westminster Station is shown in section in Figure 3. The JLE platforms are contained within 7.4 m OD (outside diameter) bored tunnels in a vertically stacked arrangement below Bridge Street.

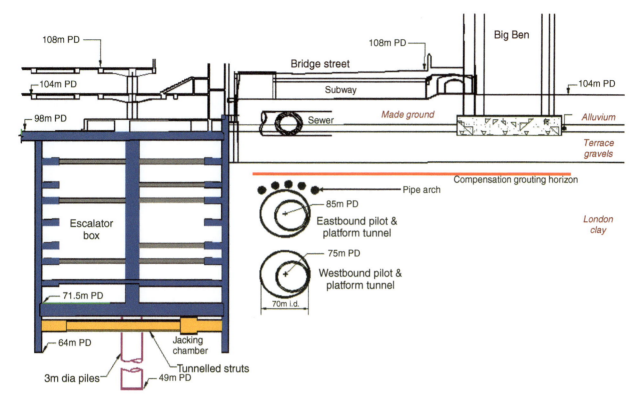

Figure 3. Section through JLE Westminster Station.

The axes of the lower westbound tunnel and upper eastbound tunnel are at depths of 30 m and 21 m below ground level respectively. Access to the platforms is by means of the 39 m deep station escalator box to the north constructed using diaphragm walls. The escalators rise to the District and Circle Line platforms and the ticket hall. The northern edge of the Clock Tower is 28 m from the centre-line of the tunnels and 34 m from the diaphragm walls of the deep escalator box. Details of the layout, the design and the construction of the station are given by Carter *et al.* (1996) and Bailey *et al.* (1999).

The station escalator box is 74 m by 28 m in plan and, with excavation up to 39 m below street level, was at that time substantially the deepest basement ever to have been constructed in London. The retaining walls consist of reinforced concrete diaphragm walls constructed to a maximum depth of 40 m from a platform 4 m below ground level. The excavation was carried out using the well known 'top-down' technique with the struts and floors being installed progressively from the top downwards as excavation progressed. In order to minimise surrounding ground movements, low-level struts were installed in tunnels close to the base of the diaphragm wall prior to excavation below the main roof slab (Crawley and Stones, 1996). The system had previously been used successfully during the construction of the Barbican Arts Centre in London (Stevens *et al.*, 1977). The location of the tunnelled struts is shown in elevation in Figure 3.

Prior to any substantial excavation within the station escalator box, the running tunnels were driven from east to west and acted as pilot tunnels for the station tunnels. These running tunnels are 4.85 m OD and were built in expanded concrete segmental linings. A Howden open-face shield with a back actor was used for both tunnel drives. An open-face shield was also used for the enlargement to form the 7.4 m OD platform tunnels. These tunnels are 160 m long and are lined with bolted segmental SGI (Spheroidal Graphite Iron) rings.

3 GROUND AND GROUNDWATER CONDITIONS

The stratigraphy at Westminster is as follows. Made Ground of depth varying between 5 m and 8 m overlies Alluvium and Terrace Gravels. The combined thickness of the Alluvium and Terrace Gravel is about 5 m. The London Clay at this location is 35 m thick and a detailed description of it is given by Burland and Hancock

(1977). The Lambeth Group below the London Clay lies over Thanet Beds which are above the Chalk. The Lambeth Group is 18 m thick. It is predominantly clayey comprising 8 m of Upper Mottled Clay dissected by a thin layer of Laminated Beds, over Lower Mottled Clay about 5 m thick. The lowermost 5 m includes a thin layer of the Pebble Bed over Glauconitic Sand. The Thanet Beds are about 8 m thick on the top of the Chalk, some 73 m below ground level.

The groundwater level in the Terrace Gravels aquifer is about 9 m below ground level with little tidal variation. Pore water pressures are close to hydrostatic equilibrium with the overlying aquifer throughout the London Clay and within the Upper Mottled Clay of the Lambeth Group. The water pressure in the Thanet Sands and Chalk is substantially reduced because of historical pumping from this aquifer.

4 ASSESSMENT OF GROUND MOVEMENTS AND POTENTIAL DAMAGE

Assessments of the differential ground movements and their impacts on the tilt of the Tower and the potential damage to the adjoining Palace of Westminster formed a vital part of the design process of the JLE Westminster Station.

The short-term ground movements associated with the tunnelling were assessed using the widely used empirical approach summarised by Attewell *et al.* (1986) and Rankin (1988) in which the settlement troughs are assumed to be Gaussian normal distribution curves. A trough width factor of 0.5 was assumed for all tunnels together with volume loss values of 2 per cent for the pilot tunnels and 3 per cent for their subsequent enlargement to form the platform tunnels. The tunnel alignment was designed to maximise the plan distance from the Clock Tower. Figure 4 shows the estimated settlement troughs for each of the individual tunnels together with the resultant tunnelling settlement trough obtained by adding the individual ones (i.e. assuming superposition to be valid). The calculated short-term settlement at the northern edge of the Clock Tower was 4.5 mm with no movement at the southern edge giving a differential settlement across the Tower of 4.5 mm. It can be seen from Figure 4 that most of the tunnelling-induced settlement is associated with the construction of the deeper westbound running and subsequent station tunnel.

Finite element analyses were used to assess the short-term movements associated with the escalator box excavation and, in particular, to assess the effectiveness of measures designed to reduce these movements.

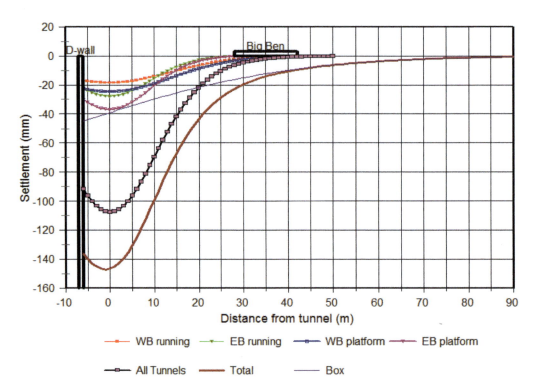

Figure 4. Estimated settlement profiles from the tunnels and the escalator box.

Of these, the most important was the inclusion of low-level struts across the box installed prior to excavation as described previously. The restraint to movement of the walls was shown to reduce the magnitude of the induced movements of the Clock Tower by up to 40 per cent and the associated differential settlements by 30 per cent. Predictions of movement were then made by applying reduction factors based on these FE results to the empirical method of Clough and O'Rourke (1990). The predicted settlement at the north and south sides of the Clock Tower due to the excavation of the escalator box were 12.7 mm and 9.8 mm respectively, giving a differential settlement of 2.9 mm across the Clock Tower.

Superposition of the estimated movements for the tunnels and the station box indicated that a maximum short-term settlement of 21.5 mm and an increase in tilt of approximately 1:2000 could be anticipated. The greater part of the settlement was expected to result from the box excavation, whereas the tunnelling was expected to produce most of the tilt of the Clock Tower.

Assessments of potential damage to the Palace of Westminster in the vicinity of the structural connection to the Clock Tower using the limiting tensile strain approach (Burland, 1995) indicated that 1:2000 was close to the maximum acceptable limit of tilt for the Clock Tower. Moreover analyses had indicated that movements could be substantially increased by interaction effects and that long-term settlements could be expected following tunnelling. It was therefore evident that additional, contingency protective measures would be needed.

5 COMPENSATION GROUTING ARRANGEMENT

As described by Harris *et al.* (2013), potential damage assessments were undertaken for all of the structures within the zone of ground movement adjacent to the proposed Westminster Station. These assessments indicated that protective measures would be required for several of them. Compensation grouting was specified as the way to protect them. The resulting proposal was to use vertical shafts to install horizontal grouting arrays above the crown levels of all the tunnels around the Westminster Station complex. The details of the shafts and grouting arrays are given by Harris *et al.* (2013).

Compensation grouting comprises the controlled injection of grout between the tunnel and the building foundations in response to observations of ground and building movements during tunnelling or excavation. As its name implies, the purpose is to compensate for ground loss. The technique requires detailed instrumentation to monitor the movements of the ground and the building. Harris (2001a) gives useful brief descriptions of various forms of grouting together with a detailed description of the one used on the Jubilee Line Extension involving fracture grouting. For this, liquid grout was injected into the London Clay from sub-horizontal steel tubes with ports at regular intervals. The ports comprise four holes spaced equally around the circumference of the tube and covered with a rubber sleeve (*manchette*). The grout is injected by inserting a probe into the tube and isolating the port to be injected by inflating packers at either side of the injection nozzle. Sufficient pressure is then applied to open the port and initiate flow into the ground. The tubes and ports are usually referred to as *tubes à manchettes* (TAMS).

Compensation grouting was adopted as the most appropriate method of ensuring that the tilt of the Clock Tower did not exceed the maximum specified value of 1:2000. The grouting arrays were extended below the full footprint of the Tower's foundations. The arrays installed below the Tower initially comprised six TAMS with a maximum spacing of 5 m. Subsequently, as a consequence of changes in other construction activities, these arrays had to be replaced and 16 TAMS were installed with maximum spacing reduced to 2.5 m. The elevation of the TAMs was constrained to be between the upper platform tunnel and the interface between the London Clay and the overlying water bearing gravels. The clay cover to the tunnel was about 5 m and the selected grouting horizon was 3 m above the crown of the upper station tunnel which was judged to give adequate cover to avoid intercepting the overlying water bearing gravels – see the vertical section in Figure 3.

6 COMPENSATION GROUTING CONTROL

The management system developed for compensation grouting was that each injection had to be defined in a grouting proposal. All injections were prescribed in terms of shaft, TAM and port number. The volume of each injection was pre-determined, and sequences of injections were specified if deemed necessary. The grouting was controlled through a monitoring control office that was also in direct communication with the tunnellers and surveyors, and had access to the real-time monitoring. Each grouting proposal incorporated communication procedures that required a positive confirmation to be received from the monitoring control office after completion of a specified part of the proposal before work could proceed. For example, with an advancing tunnel, a pattern of injections relative to the face was defined (Harris *et al.*, 1996). Injection could only start once a given stage of the tunnelling cycle had been completed. Moreover the injections would have to be completed before tunnelling could proceed beyond a specific point.

To maintain the flexibility needed to modify the grouting proposals in response to observed behaviour, frequent meetings were held to review construction progress, grouting records and monitoring results. Minor modifications were made by omitting injections or changing grout volumes. If significant changes were necessary, a revised proposal would be produced. The short time-scales for production, discussion, amendment and consenting to grouting proposals, required a co-operative approach from all parties to avoid delays to the works. Further details of the strategy adopted for implementing the compensation grouting are given by Harris *et al.* (2000) and Harris (2001b).

7 INSTRUMENTATION

Successful implementation of the compensation grouting technique requires reliable and accurate monitoring, its rapid processing and dissemination and informed interpretation in conjunction with records of construction activities. The following instrumentation was installed on the Big Ben Clock Tower.

7.1 *Tilt monitoring*

Tilt of the Clock Tower was identified as the most important parameter to monitor and consequently a range of independent systems were used.

An optical plumb had been used to monitor the tilt of the Clock Tower during the construction of the New Palace Yard underground car park in the early 1970s (Burland and Hancock, 1977) and had been read intermittently over the intervening period. The original target which was removable was still available and it was decided that the JLE surveyors should monitor this using a Wild ZL optical plumb. A new datum was established which was related to the original datum giving a self-consistent data set extending over a period of nearly thirty years. A re-designed target was procured to improve the repeatability of the readings. Observations were taken on each of four faces in the north-south and east-west directions and then averaged to give a reading. The observations were recorded to a resolution of 0.1 mm and the four independent observations generally lay within a range of 0.5 mm. The resolution and accuracy over the 55.4 m gauge length are equivalent to tilts of about 1:550,000 and 1:110,000 respectively.

Retro reflective prism targets attached to the north, east and west clock faces were surveyed to give displacements in three dimensions using a Leica TC1610 Total Station. The observation of each target required a separate set-up location and hence the three measurements were entirely independent. Readings were taken to a resolution of 0.1 mm for distance and 1″ of arc on horizontal and vertical angles. The repeatability of the readings was ±2 mm or 1:13,000.

In order to avoid the need for excessive surveying resources during grouting episodes, a real-time monitoring system was necessary. Eight electrolevels were installed on 1 m long beams, six of which were mounted horizontally and the other two vertically. Four were oriented to measure north-south tilt and four to measure east-west tilt. In the event, these instruments were not used to control the works due to the success of an alternative real-time system developed by the contractor – an electronically monitored plumb line.

The electronically monitored plumb line was developed by the contractor's surveying department and was named the "Gedometer" after its primary creator Gerald "Ged" Selwood. The instrument comprised an invar strip suspended from a grillage in the belfry over a ventilation shaft which extends over the full height of the Tower in its north-east corner. A temporary decking was installed in this shaft 5 m above ground level (which gave an almost identical gauge length as the optical plumb) on which a digitising tablet was installed. A puck

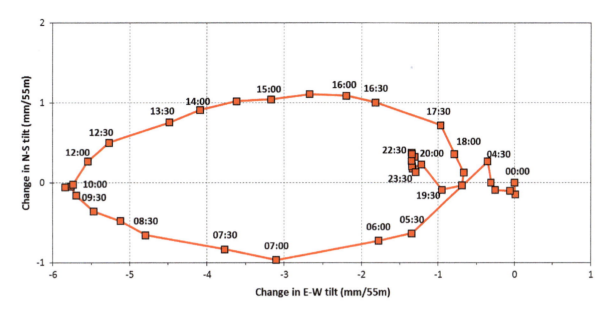

Figure 5. Measured horizontal movements at the height of the clock face taken at half-hourly intervals during a single day (18th June, 1996).

was suspended from the end of the plumb line and the tablet programmed to automatically record the location of the puck at 1 second intervals. Individual observations were averaged to produce a reading at specified intervals – generally 30 minutes was found to be adequate. The instrument performed reliably over a period of 4½ years with only occasional adjustment and maintenance. In these instances corrections to the recorded movements have been necessary. The readings were reported to 0.001 mm and the accuracy was arguably as good as 0.1 mm or 1:550,000 over the period of individual compensation grouting episodes.

It is important to appreciate that a structure such as the Clock Tower undergoes significant movements due to daily and seasonal temperature changes. Burland and Hancock (1977) noted that there is a seasonal east-west movement of about 6 mm at a height of 55 m with the Clock Tower moving westwards in the summer and eastwards in the winter. The measurements made with the "Gedometer" recorded a maximum daily range of tilt of 6.2 mm in the east-west direction compared to a cycle of about 2.5 mm in the north-south direction. The magnitude of the thermal bending is related to the number of hours of sunshine during the day. Consequently it is at its greatest in the summer months, when the top of the Tower traces an approximately elliptical path with its major axis in the east-west direction as shown in Figure 5. Burland and Viggiani (1994) report very similar daily movements of the Pisa Tower. An important consequence of this seasonal and daily behaviour is that it is difficult to determine reliably small long-term trends in changes of inclination of the Clock Tower. This important topic is discussed later.

As described by Harris *et al.* (2013) comprehensive precision levelling was carried out in the adjoining Palace of Westminster. Figure 6 shows the location of the 42 levelling points in relation to the Big Ben Clock Tower. Four levelling points on the corners of the Tower were used to calculate the tilt of the Tower. The accuracy of the levelling of these four points was generally ±0.3 mm and their spacing about 12 m. This gives an accuracy in the measurement of tilt of about 1:20,000. Four temperature sensors were installed in the Clock Tower.

8 MEASURED TILT OF THE CLOCK TOWER DURING CONSTRUCTION

Monitoring of the Clock Tower for tilt began in November 1994. Monitoring was carried out throughout construction and continued for 15 years after construction ended. Figure 7 shows the measured North-South tilts of the Clock Tower from the optical plumb throughout the construction period and for six months thereafter. The timings of the main construction activities are also shown in the figure. The passage of the four tunnel

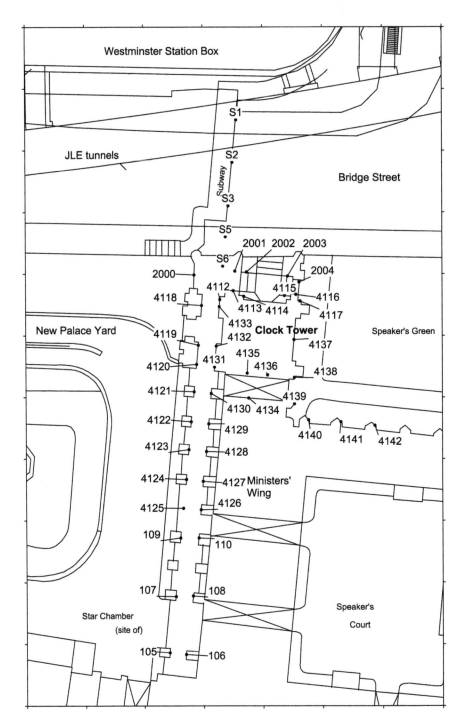

Figure 6. Plan showing location of precision levelling points. Points 4112, 4131, 4138 and 4117 were used to measure the tilt of the Clock Tower. Point 106 was taken as the datum for monitoring the Clock Tower.

drives are shown across the top of the figure and the timing at various excavation depths in the escalator box are shown across the bottom.

The thick vertical line at December 1995 indicates the start of compensation grouting to control the tilt of the Clock Tower and the various episodes of grouting are shown across the middle of the figure, the final episode being in September 1997.

Initially performance control levels (PCL) of tilt were set conservatively at 1:6,000 and 1:4,000 for the amber and red triggers. These limits are equivalent to about 9 mm and 14 mm relative northward movement at the height of the clock face.

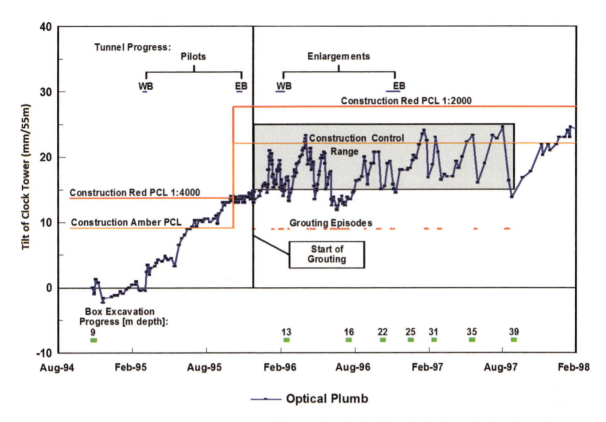

Figure 7. Optical plumb measurements of the Northward tilt of the Clock Tower during the works.

As predicted, the first activity to affect the tilt of the Clock Tower was the westbound running tunnel drive, which was undertaken in March 1995. A tilt to the north of about 4 mm was observed as shown in Figure 7. The northward tilt continued to increase significantly following the tunnel drive and reached the amber PCL of 9 mm in July 1995. This initiated a reappraisal of the potential movements. It was also evident that it was necessary to minimise further movements and to develop a strategy for implementing compensation grouting.

Back-analysis of the observed settlements associated with the westbound tunnel indicated that the volume losses in this area were about 3 per cent – significantly larger than had been allowed for in the design settlement assessments. The reasons for this are discussed by Standing and Burland (2006). The observed increase in tilt after the westbound tunnel drive also demonstrated that substantial time-dependent movements should be expected both during the construction period and subsequently. The following actions, *inter alia*, were undertaken.

1. Co-ordination of tunnel advance with implementation of grouting to allow settlements to be fully compensated; this was referred to as *concurrent* compensation grouting.
2. The red PCL on the permissible increase in the northward tilt of the Clock Tower was raised to 1:2,000.
3. The amber PCL on the tilt of the Clock Tower at which grouting would be instigated was raised to 1:2,500.
4. A trial grouting episode below the Clock Tower was undertaken to demonstrate that control of the tilt of the Tower could be exercised.
5. An expert review panel was set up to advise on geotechnical and construction issues relating to the Clock Tower.
6. Close liaison with the Parliamentary Works Directorate was maintained through its geotechnical advisor.
7. The finite element analysis undertaken at the design stage was updated to take account of revised construction methods and sequences. The analysis was calibrated against the observed settlements to give the best possible prediction of future movements. The results of this finite element analysis assisted in identifying potential mechanisms of movement of the Clock Tower and the adjoining Palace of Westminster and allowed variations in the excavation and construction procedure of the station escalator box to be investigated.

Because the Clock Tower was outside the zone of influence of the shallower eastbound tunnel, its drive was permitted to go ahead in October 1995 before the trial grouting below the Clock Tower mentioned in item 4 above. As can be seen from Figure 7 no noticeable increase in tilt took place at this time.

The trial grouting was carried out in December 1995 at which time the northward tilt was 14 mm. The trial was inconclusive and a further trial was delayed until February 1996 due the necessity of installing new TAMs below the Clock Tower. During the second trial the tilt was reduced by about 5 mm confirming both the suitability of the method and that significant control could be exercised.

The next grouting episode below the Clock Tower was concurrent grouting associated with the enlargement to form the westbound station tunnel. Grouting within the settlement trough was fully coordinated with tunnel advances and was augmented by additional injections below the Clock Tower. The aim was to produce full compensation for the tunnelling-induced settlements together with a small reduction in tilt. This was successfully achieved, with the tilt being reduced by 5 mm. Thereafter, during excavation of the escalator box from 9 m depth to 39 m and the enlargement for the eastbound platform tunnel, grouting was undertaken in response to the observed tilts of the Clock Tower rather than being directly related to construction activities. A construction control range of 15 mm to 25 mm tilt was adopted. It can be seen from Figure 7 that the upper limit of this control range was not exceeded throughout the construction period although occasionally the lower limit was exceeded. In total, 24 episodes of grouting were undertaken between January 1996 and September 1997 over which period a total volume of 122 m^3 of grout were injected beneath the Clock Tower. In general, grouting was confined to the northern half of the raft foundation as shown in Figure 8 for a typical grouting episode. Harris (2001b) gives a more detailed account of the grouting procedures adopted.

9 SETTLEMENT MONITORING DURING CONSTRUCTION

The locations of the monitoring points are shown in Figure 6 and point 106 was usually taken as the datum. Points 4112 and 4131 are located on the Clock Tower itself. Figure 9 shows the measured settlements along the west façade of the Palace of Westminster at various stages. Note that points 4112 southwards to 106 are located on the Palace of Westminster as shown in Figure 6. Also shown in Figure 9, for the purposes of comparison, is the resultant estimated settlement trough with no compensation grouting taken from Figure 4.

The settlement observations plotted in Figure 9 correspond to immediately before and immediately after each of the four main tunnel drives and from the end of construction in September 1997.

The following observations relate to the settlements of the western façade of the Palace:

1. Settlements extend southwards to a distance of 30 m from the north face of the Clock Tower (i.e. from survey point 4112).
2. The shape of the measured settlement profile is similar to the resultant estimated settlement trough but the magnitude is substantially less.
3. Even though the compensation grouting only extended beneath the northern half of the Clock Tower its effects extended southwards as far as point 4127 (see Figure 6) which is about 25 m from the plan extent of the grout injections.

10 POST-CONSTRUCTION BEHAVIOUR

Following the completion of construction in September 1997 the tilt of the Clock Tower continued to increase because of the ongoing consolidation of the London Clay. By the end of 1997 it was evident that, although the rate of tilting was clearly decreasing, the agreed Performance Control Level (PCL) of 1:2,000 would be exceeded within a matter of months. The compensation grouting facilities were still in place and could be used if necessary. However it was felt that a further episode of grouting might accelerate the movements again and be counterproductive. It was therefore decided to review the PCL in the light of the monitored behaviour to ascertain whether it could be relaxed. A thorough review of the historical monitoring data, background movements, performance during construction and available long-term settlement data from the JLE elsewhere in the Westminster area was undertaken (Harris, 2001b). A prediction of the probable increase in tilt was made

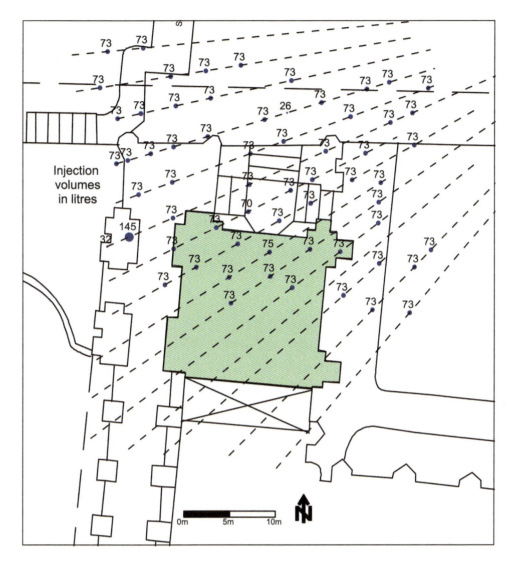

Figure 8. Typical grout injection pattern.

in March 1998, 6 months after the end of construction and the final compensation grouting episode. This suggested that the tilt would increase to about 40 mm over a period of about 5 years.

Examination of the crack width measurements on the Palace of Westminster revealed the mechanism by which an increase in the tilt of the Clock Tower was reflected in increased crack widths within the structure. A detailed statistical analysis that took account of the effects of temperature change showed that for each 1 mm increase in tilt over the vertical gauge length of 55 m the average crack width at a high level in the building would increase by 0.07 mm. This correlation together with the predicted increase in tilt suggested an increase in crack width of just under 3 mm. An assessment of the form of the structure and of the existing cracking led to the conclusion that an opening of the existing cracks by up to 3 mm would not significantly affect the ease and cost of repair (i.e. the level of damage – see Burland, 1995). On this basis it was agreed that the PCL on tilt could be raised to about 40 mm.

Figure 10 shows the results obtained from the optical plumb measurements over the period of time since they were initiated in the early 1970s for the construction of the underground car park (Burland and Hancock, 1977). As reported by these authors, the construction of the car park caused the tower to tilt about 2.5 mm to the south at a height of 55 m. The measurements show that in the 22 years following the completion of the car park the Clock Tower underwent a background rate of tilt to the north of about 0.65 mm per annum. The fluctuations around this trend of about ±2.5 mm are due to the seasonal and daily thermal effects mentioned previously.

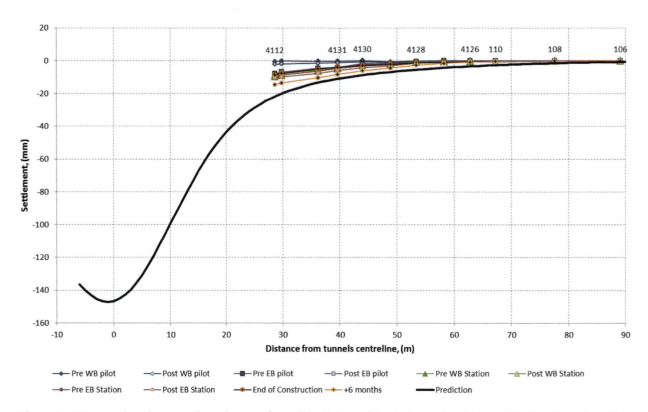

Figure 9. Measured settlements along the west face of the Palace of Westminster (see Figure 6 for location of levelling points).

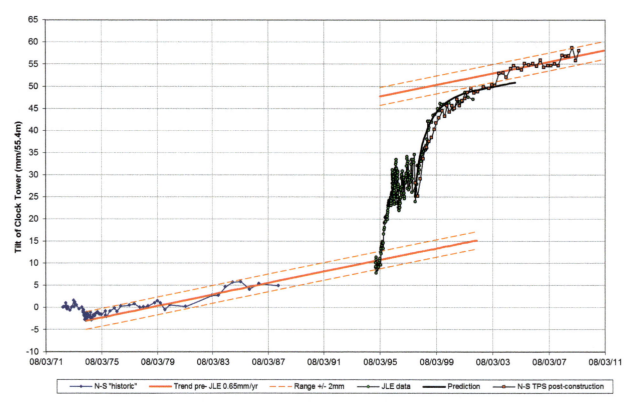

Figure 10. The long-term northward movements of the Clock Tower prior to and subsequent to the construction of the Jubilee Line Extension underground railway.

Figure 10 also summarises the changes in north-south tilt of the Clock Tower during and subsequent to the construction of the new JLE station. It can be seen that, after the last grouting episode in September 1997, the trend has been a reasonably steady reduction in the rate of northerly tilt. Since about April 2003 the rate of tilt has reduced to the background rate of about 0.65 mm per annum established over a period of 22 years prior to the construction of the JLE. Taking account of this background rate it would appear that the magnitude of the tilt induced by the construction activities amounted to about 38 mm at the height of the clock face. It has been estimated that, had compensation grouting not been carried out, the construction-induced northward tilt would have been about 130 mm which would have caused significant damage to the Palace of Westminster.

11 DISCUSSION AND CONCLUSIONS

This chapter describes the geotechnical measures that were used to protect the Big Ben Clock Tower from ground movements resulting from the construction of the nearby Jubilee Line Extension underground tunnels and the new Westminster Station.

The key conclusions that can be drawn from the work are as follows:

1. High-quality real-time measurements, particularly of the tilt of the Clock Tower, were vital to understanding its behaviour, to demonstrating the need for protective measures, and to controlling the compensation grouting that was implemented.
2. Before and during the works valuable insights into the effects of the construction works and possible responses of the buildings were gained from careful interpretation of appropriate numerical modelling.
3. The compensation grouting works were strictly regulated through a monitoring control office and proved extremely successful in controlling the settlement and tilt with considerable precision.
4. The understanding gained during the works led to an effective predictive model with which to estimate and monitor the time-dependent tilt of the Clock Tower.
5. More generally, while numerical modelling can be helpful in understanding mechanisms of behaviour, it is not possible to make precise predictions of the form and magnitude of ground movements due to excavation and tunnelling. These depend on a number of factors that are usually not known at the design stage, the most important of which are the detailed construction method and timing and their influence on the ground properties. Therefore it is essential to monitor the movements as they develop and thereby progressively refine understanding of both the ground movements and the response of the nearby structures impacted by them.
6. In this case the finite element analysis undertaken at the design stage was updated to take account of the revised construction methods and sequences. This revised analysis was calibrated against the observed settlements to give the best possible prediction of future movements.
7. If there is a risk of unacceptable damage due to ground movements, mitigation measures should be designed and put in place at the beginning of the work so that they can be implemented speedily and with a minimum of disruption.
8. For historic buildings the level of damage and the mitigation methods used to control it should be in accordance with the principles of conservation as far as is practicable.

REFERENCES

Attewell, P.B., Yeates, J. and Selby, A.R. (1986). *Soil movements induced by tunnelling and their effects on pipelines and structures.* Blackie, Glasgow.

Burland, J.B., Standing, J.R., Linney, L.F., Mair, R.J. and Jardine, F.M. (1996). A collaborative research programme on subsidence damage to buildings: prediction, protection and repair. In: Mair, R.J. & Taylor, R.N. (eds.) *Geotechnical Aspects of Underground Construction in Soft Ground*, Balkema, Rotterdam, pp. 773–778.

Burland, J.B. (1995). Invited Special Lecture: Assessment of risk of damage to buildings due to tunnelling and excavation. In: *1st Int. Conf. on Earthquake Geotechnical Engineering*, Tokyo, 3, pp. 1189–1201.

Burland, J.B. and Viggiani, C. (1994). Osservazioni sul comportamento della Torre di Pisa. *Rivista Italiana di Geotecnica*, 28 (3), pp 179–200.

Burland, J.B. and Hancock, R.J.R. (1977). Underground car park at the House of Commons, London: geotechnical aspects. In: *The Structural Engineer*, 55 (2), pp. 87–100.

Bailey, R.P., Harris, D.I. and Jenkins, M.M. (1999). Design and Construction of Westminster station on the Jubilee Line Extension. In: *Proc. Instn. Civ. Engng, Jubilee Line Extension* 1999, **132**, pp. 36–46.

Carter, M.D., Bailey, R.P. and Dawson, M.P. (1996). Jubilee Line Extension, Westminster Station design. In: Mair, R.J. & Taylor, R.N. (eds.) *Geotechnical Aspects of Underground Construction in Soft Ground,* Balkema, Rotterdam, pp. 81–86.

Clough, G.W. and O'Rourke, T.D. (1990). Construction induced movements of insitu walls. In: *Design and performance of earth retaining structures – Proc 1990 Speciality Conf.* Geotechnical Special Publication 25, ASCE, New York, pp. 81–86.

Crawley, J.D. and Stones, C.S. (1996). Westminster Station – Deep foundations and top down construction in central London. In: Mair, R.J. & Taylor, R.N. (eds.) *Geotechnical Aspects of Underground Construction in Soft Ground.* Balkema, Rotterdam, pp. 93–98.

Harris, D.I., Mair, R.J., Burland, J.B. and Standing, J.R. (2013). Protective compensation grouting beneath the Big Ben Clock Tower. In: Bilotta, E., Flora, A., Lirer, S. and Viggiani, C. (eds.) *Geotechnics and Heritage.* Taylor and Francis Group, London, pp. 137–152.

Harris, D.I. (2001a). Chapter 11, Protective measures. In: Burland, J.B., Standing, J.R. and Jardine, F.M. (eds.): *Building response to tunnelling. Case studies from the Jubilee Line Extension, London*, Volume 1, *Projects and methods.* CIRIA Special Publication 200. CIRIA and Thomas Telford, London, pp. 135–176.

Harris, D.I. (2001b). Chapter 18, The Big Ben Clock Tower and the Palace of Westminster. In: Burland, J.B., Standing, J.R. and Jardine, F.M. (eds.): *Building response to tunnelling. Case studies from the Jubilee Line Extension, London*, Volume 2, *Case Studies.* CIRIA Special Publication 200. CIRIA and Thomas Telford, London, pp. 453–508.

Harris, D.I., Mair, R.J., Burland, J.B. and Standing, J.R. (2000). Compensation grouting to control tilt of Big Ben Clock Tower. In: Kusakabe, O., Fujita, K. & Miyazaki, Y. (eds.): *Geotechnical Aspects of Underground Construction in Soft Ground.*, Balkema, pp. 225–232.

Harris, D.I., Pooley, A.J., Menkiti, C.O. and Stephenson, J.A. (1996). Construction of low-level tunnels below Waterloo Station with compensation grouting for the Jubilee Line Extension. In: Mair, R.J. & Taylor, R.N. (eds.) *Geotechnical Aspects of Underground Construction in Soft Ground.* Balkema, Rotterdam, pp. 361–366.

Rankin, W.J. (1988). Ground movements resulting from urban tunnelling; predictions and effects. *Engineering Geology of Underground Movement,* Geological Society, Engineering Geology Special Publication No. 5, pp. 79–92.

Standing, J.R. and Burland, J.B. (2006). Unexpected tunnelling volume losses in the Westminster area, London. *Geotechnique,* **56**, No. 1, pp. 11–26.

Stevens, A., Corbett, B.O. and Steele, A.J. (1977). Barbican Arts Centre: the design and construction of the substructure. *The Structural Engineer,* Vol. **55**, pp. 473–485.

Geotechnics and Heritage: Historic Towers – Lancellotta, Flora & Viggiani
© 2018 Taylor & Francis Group, London, ISBN 978-1-138-03272-9

Preservation of historic towers in Venice: The instructive monitoring-driven intervention on the foundations of the Frari bell tower

G. Gottardi & M. Marchi
Department of Civil, Chemical, Environmental, and Materials Engineering, University of Bologna, Italy

A. Lionello
Ministry for Cultural Heritage and Activities, Italy

C. Rossi
R.TEKNOS SrL, Bergamo, Italy

ABSTRACT: Historical documents report that Venice used to have more than one hundred bell towers. The peculiar environmental conditions in which foundations had to be built, however, have typically generated over the centuries substantial differential settlements and potential instability of such structures. Also thanks to the advanced skill reached in the foundation construction technique, about 80 towers are still presently standing, obviously requiring an increasing attention and awareness from public authorities of their specific vulnerability. Special control and maintenance programs have been thus devoted to bell towers and preservation interventions have been carried out in full respect of their historic, architectural and structural features. An emblematic example of such an approach is represented by the multiphase intervention carried out on the bell tower of the Frari Basilica, the second tallest in Venice. The tower was affected by a slow but constant differential settlement with respect to the adjacent masonry structures of the Basilica. Starting some 15 years ago, modern remedial measures were therefore implemented, first on the foundations – using a ground improvement intervention by fracture grouting – and then on the elevation structure. This chapter aims at presenting the well-documented Frari case study, from the preliminary crucial and accurate site investigations to highlighting both the strategy of improving the overall safety – without altering the original structure and without substantially modifying the current stress distribution – and the innovative methodology, adopted throughout, of a gradual and modular design, constantly driven by the outcome of an extensive real-time monitoring system of the soil-structure interaction.

1 INTRODUCTION

Historic towers and bell towers are delicate structures, a fundamental component of the local historic and architectural Venetian heritage. Nevertheless, they have often been neglected in the past. Historical documents report that Venice used to have more than one hundred towers (Fig. 1): a few were demolished, others collapsed for various reasons (like foundation settlements, lightning, fires, etc.). To date, 85 are still standing, thus playing their role of distinctive architectural feature of the Venetian skyline.

The *San Marco* bell tower, standing in the central San Marco square, is famous worldwide and, with its 98.6 m is the tallest of the city. The shortest is *Sant'Eufemia* while *San Samuele* is the oldest. The inclinations of *Santo Stefano* and *San Giorgio dei Greci* are impressive and remind of their possible typical trend, made well-known by the Pisa Tower. Many towers have been demolished or have collapsed over the centuries and sometimes their remains can be seen around the city. This is the case of *Santa Margherita* (drawing K in Fig. 1), which was partly demolished in 1808 for safety reasons. In other cases, like *Sant'Agnese,* the remains are no longer visible, but the notes collected by a far-sighted technician after the tower demolition have brought us a typical and very instructive example of a Venetian bell tower foundation (Fig. 2). Such historical information on traditional construction methodologies becomes very useful to deduce the missing data of each single case and to confirm the results of specific surveys. The structural analysis of historic buildings, such as ancient bell

Figure 1. Typical Venetian bell towers: 14 towers (A: *S. Luca*, B: *Ognissanti*, C: *S. Fantino*, D: *S. Gio. Elemosinario*, E: *S. Maria Maggiore*, F: *S. Eustachio*, G: *S. Ubaldo*, H: *S. Gallo*, I: *S. Angelo di Marzorbo*, K: *S. Margherita*, L: *S. M. Della Presentazione*, M: *S. Clemente*, N: *S. M. del Pianto*, O: *S. M. Madd. Delle Penitenti*) selected from a collection of 104 drawings (modified from Levi, 1890).

towers in Venice, is, in fact, a path often full of uncertainties. The peculiar environmental conditions in which foundations of historic buildings had to be built in Venice produced a typical and recurrent structural form. Scarce availability of materials, water table at the ground level and very low bearing capacity of the immediate subsoil imposed severe constraints. The fabric of most foundations remained substantially unchanged until the beginning of the last century, with two predominating types: shallow masonry foundations (for low buildings, not bordered by canals) and wooden piled foundations (for major buildings and usually for walls bordering canals) (Zuccolo, 1975).

Because of the shallow layer of lagoon soft silty clay, especially for major building, typical settlements up to 40–50 cm tend to develop. Related differential settlements for towers produced unavoidable damaging interactions with the adjacent structures and severe consequences on their own stability.

Nevertheless, in the past centuries local authorities often neglected the bell tower maintenance: churches and other main buildings had priority. This situation went on until the beginning of the last century, when the collapse of the *San Marco* bell tower in 1902, the latest in a long list but the most important though, caused major concern and developed a new awareness for the preservation of these monuments not only in the Venetian municipality but all over the Italian peninsula.

Immediately after the collapse, in 1902, several studies and monitoring activities started to evaluate the stability conditions – and consequent possible remedial interventions – on many Venetian towers considered at stake, including *Santo Stefano* and Frari bell towers. A singular example of remedial measure suggested to stop the evolution of *Santo Stefano* inclination is shown in Figure 3 (left).

The intervention was eventually carried out from 1903 to 1905, based on a different design (right picture in Fig. 3). It was characterized by the construction of five masonry buttresses located on the leaning side along

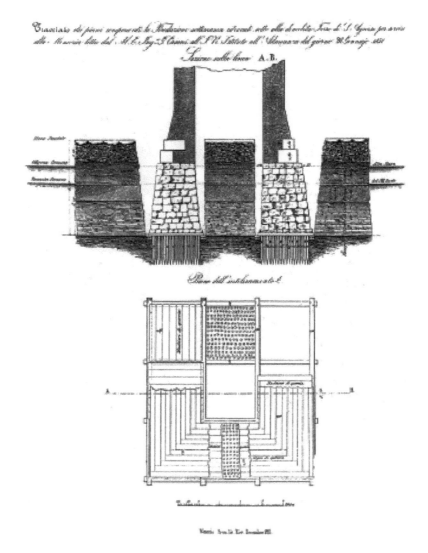

Figure 2. Section and plan of the foundation of *Sant'Agnese* bell tower, demolished in 1851 (from Casoni, 1851).

the adjacent canal, resting on a rectangular concrete base (4 m × 10 m) and founded on 3 m long concrete screw piles. A recently installed monitoring system would suggest on the one hand the effectiveness of the intervention, but on the other hand the constant need of suitable controls.

The intervention on the Frari bell tower will be the subject of the next sections of this chapter. Finally, it is worth noticing that such more recent attention for towers is currently being kept constantly high and their particular vulnerability is now carefully considered by special control and maintenance programs. In such a framework, the local authority in charge of monuments and historic buildings preservation (*Sopraintendenza per i beni architettonici e paesaggistici di Venezia e laguna*) supported the implementation of an important project with the purpose of gathering as much relevant information on bell towers as possible in order to devise a reference database made up of 90 monographic tables (Lionello, 2011). On the basis of such a database and the analysis of relevant safety conditions it is now possible to establish the possible intervention priorities.

2 THE FRARI BELL TOWER: A HISTORY OF INTERVENTIONS

The Basilica of *Santa Maria Gloriosa dei* Frari is one of the largest and most splendid churches in Venice. It stands on the *Campo dei* Frari at the heart of the city (Fig. 4). Historical archives tell us that the Franciscans

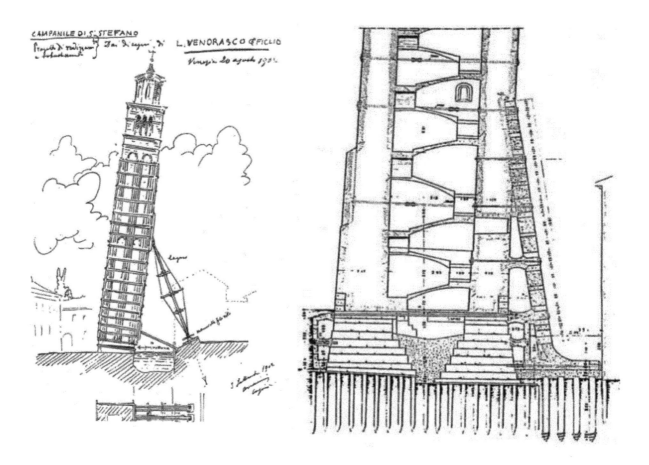

Figure 3. Left: Sketch of a strengthening intervention suggested for the Santo Stefano bell tower in 1902 (Photo archive of the "*Soprintendenza per i beni architettonici e paesaggistici di Venezia e laguna*"); Right: the intervention with counterfort actually carried out in 1905.

were initially granted land to build a church in 1250, but the first building was not completed until 1338. Works almost immediately (in 1340) started again on its much larger replacement, the current church, which took over a century to be built. The Frari bell tower (Fig. 4), the second tallest in the city after that of *San Marco*, was built between 1361 and 1396. The bell tower structure is 9.5 m wide at the ground level, 65 m tall and weighs about 57 MN. The internal ramp staircase up to the belfry is supported by a double structure of thick brick masonry. It was originally conceived as a fully independent structure, but during the reconstruction of the Basilica, the bell tower was included into the masonry walls, at the south-east corner of transept and left aisle (see plan of Fig. 4). The connection between the bell tower and the Basilica is at the origin of the subsequent problems that have always affected their structures. In 1432 the St. Peter's chapel was constructed adjacent to the Basilica and the bell tower, even if structurally independent. The first documented signs of deterioration of the structures date back to the end of 16th century. Between the end of the 19th century and the first decade of the 20th, three main strengthening interventions, widely documented by projects, surveys and site sketches, were carried out (Lionello, 2008).

In fact, the bell tower differential settlement had caused, over the centuries, major damages both to the St. Peter's chapel and to the vaults of the left aisle of the church, requiring urgent repair interventions on the masonry walls and at the level of the foundations. In the first years of the 20th century the tower showed a differential settlement of already about 0.40 m with respect to the Basilica and an out of plumb toward south-east of 0.765 m at a height of 42.5 m (resulting in an inclination of about 1°). Only after these repair works were the structures of the bell tower and the St. Peter's chapel connected to each other at the foundation level, with a solid relieving arch made by three brick layers, and at the superstructure level, by metal ties. In 1904, following the universal concern after the sudden collapse of the world famous St. Mark bell tower, an extensive investigation was carried out on many Venetian slender structures considered at stake. In the Frari

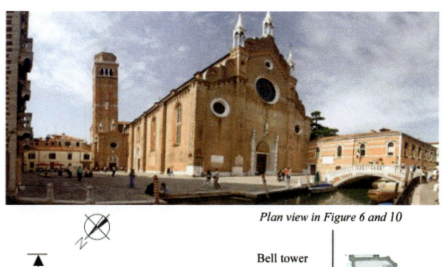

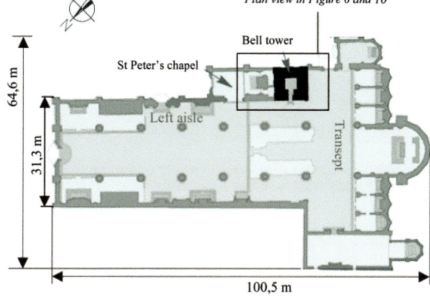

Figure 4. Top: a view of the Frari Square (*Campo dei* Frari) and its monuments: the bell tower and the basilica. Bottom: a plan of the basilica with the bell tower and the St. Peter's chapel.

case, the surveys revealed inadequate foundations with respect to the bulk of the bell tower, this being therefore the main cause of the tower settlements. For this reason, a strengthening intervention on the bell tower raft foundations was carried out, consisting of widening its base, starting from the south side (toward which the tower was essentially leaning). At the time, the intervention was designed according to the traditional Venetian soil strengthening technique (Marchi *et al.*, 2006), with the insertion of closely spaced timber piles. The piles, 3.80 m long, 0.20 × 0.20 m of transverse dimensions and essentially touching each other, were made from larch and covered by a 2.00 m wide concrete bed, parallel to the side of the bell tower (Fig. 5 and 6).

All the masonry walls were also treated with cement mortar to restore the cohesion of the brick masonry and improve its strength. Such raft foundation enlargement had been most probably designed as to be extended to the other sides; as a matter of fact, however, for emergency reasons that gave priority to other interventions, it was never completed (Lionello *et al.*, 2007).

3 INVESTIGATIONS AND MONITORING BEFORE THE INTERVENTION

An extensive survey of the general safety conditions of Venice bell towers was planned and implemented at the beginning of the 1990s by the local Architectural Heritage Office of the Italian Ministry for Cultural Heritage and Activities. Afterwards, a rather detailed diagnostic investigation of the Frari bell tower started, including photogrammetric survey, crack-pattern survey carried out with the aid of climbing technicians, endoscopies,

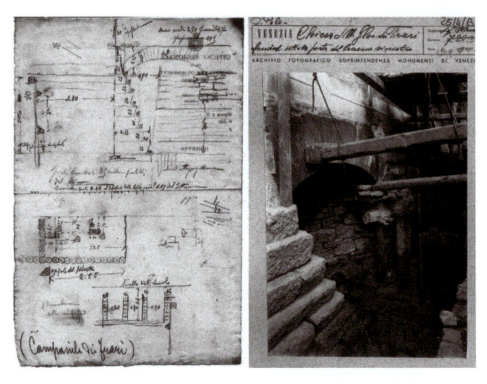

Figure 5. Right: picture of the strengthening intervention on the bell tower foundations, dated to 1904. Left: sketch of the foundations drawn during the works (Photo archive of the "*Soprintendenza per i beni architettonici e paesaggistici di Venezia e laguna*").

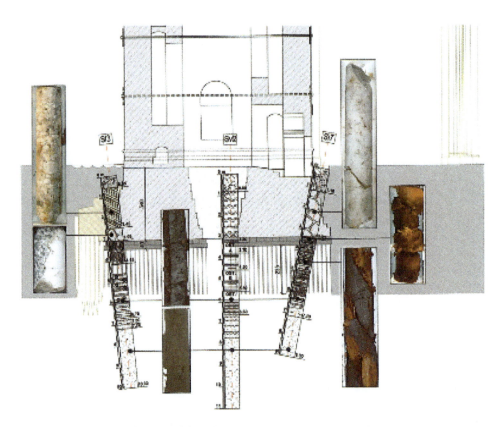

Figure 6. Reconstructed section of the Frari foundation with the enlargement of the 1904 on the left and pictures of samples extracted during the recent site investigation (Lionello, 2008).

Figure 7. Possible evolution of the bell tower displacements after the bell tower foundation enlargement in 1904 and interaction with the adjacent basilica structures (Lionello *et al.*, 2004).

single and double flat-jack tests on the masonry walls, sonic tests on steel ties, monitoring of the main cracks by means of extensometers and installation of clinometers for the detection of rotations of the bell tower, as well as preliminary geotechnical investigations of the foundation soil.

In 2000, a further survey aimed at detecting the differential settlements of the bell tower, by comparison with the 1902 situation, revealed settlements as large as 49.8 mm in the east corner, 61.3 mm in the south corner and 92.3 mm in the west corner of the bell tower base (Lionello *et al.*, 2004). These findings, together with the results of the photogrammetric survey of the 1990s, indicated that the bell tower was tilting in the opposite direction with respect to the "historical" trend, essentially moving back toward its vertical position (Fig. 7). This was most probably the consequence of the 1904 foundation widening.

Emergency interventions were provided to the structures more affected by the deformation processes and a monthly survey carried out from 2000 to 2004 confirmed the displacement trend, recording an average settlement of about 1 mm/year in the west corner of the tower.

3.1 *Measurement of the state of stress*

Flat-jack testing technique was used to measure the existing state of stress on the masonry structures of the bell tower and of the adjacent Basilica.

At the base of the bell tower an average value of 1.92 MPa was measured on the outer side, while on the inner part a mean value of 1.44 MPa was estimated (Fig. 8).

Inside the bell tower, flat-jack tests were carried out on both sides of an inclined oblique crack which runs along the south-west side of the tower. The similar stress values measured at two different levels, clearly show that the crack, even if passing through the entire thickness of the wall, does not induce particular stress concentration on the structure. Very high values of compressive stress were measured at the top of the column sustaining the propped arch: 1.76 MPa on the outer side and 3.20 and 3.04 MPa on the inner side (Fig. 8). A detailed analysis was also carried out on the wall of the Basilica adjacent to the bell tower and the results indicated the presence of a thrust line going from the bell tower to the structures of the Basilica (shown in Fig. 4), with stress values ranging from 0.56 MPa to 0.95 MPa. The presence of this thrust line was also

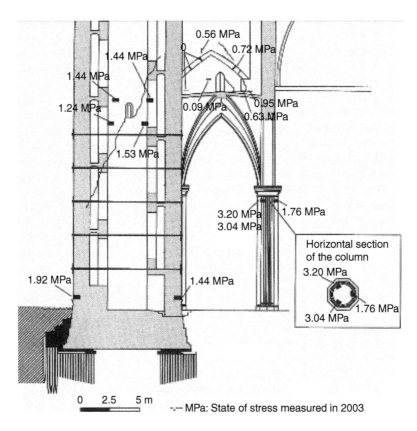

Figure 8. Values of the state of stress measured by flat-jack test in the structures of the Frari basilica and of the bell tower (Lionello, 2008).

confirmed by the results of the relevant numerical model, carried out during the investigations (Lionello, 2008).

3.2 *Geotechnical investigations*

A rather detailed geotechnical investigation was carried out in May 2003 (tests "x"/A in Fig. 9) and, on that basis, the subsequent ground improvement intervention by soil fracturing was planned.

The investigations of such A-phase consisted of: 4 piezocone tests, 2 vertical and 5 inclined continuous coring boreholes, 4 continuous borings into the foundation block, together with the extraction of several undisturbed soil and foundation block samples for the subsequent execution of the laboratory tests, which enabled the stratigraphy, the subsoil properties and the geometry of the foundation block to be defined with some detail (Gottardi *et al.*, 2009). A section of the peculiar bell tower foundation along the SE-NW direction is provided in Figure 6 and Figure 10.

As usual in Venice (Marchi *et al.*, 2006), the Frari bell tower foundation is made up of Istrian limestone squared blocks and short timber compaction piles (1.70–2.20 m long), with an interposed 0.40–0.50 m thick larch boarding.

The soil profile under the tower, shown in Figure 10, consists of:

– Unit A, between ground level and a depth of about 3.2 m: anthropic fill;
– Unit B, between 3.2 m and about 6.7 m: dark grey, soft, silty clay, with occasional organic material, normally consolidated or slightly overconsolidated, with organic inclusions and shells. In Figure 11 the relevant main geotechnical characteristics, as deduced from laboratory and in situ tests, are reported;
– Unit C, between 6.7 m and 14 m: grey medium-fine sand, non plastic, from dense to very dense;
– between 14 m and the maximum investigated depth: alternation of soft clayey silt and medium-fine dense sand.

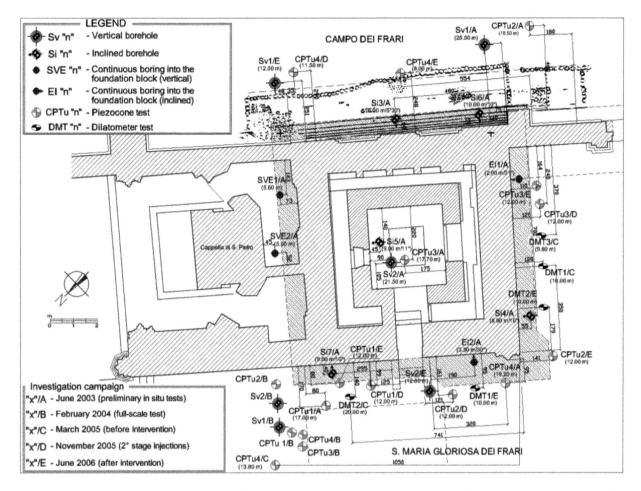

Figure 9. Plan of the Frari bell tower with the location of the geotechnical in situ investigations at various stages of the intervention (Gottardi et al., 2013).

Finally, the good conditions of the foundation enlargement built on the external side of the bell tower were confirmed. A careful evaluation of settlement trend with time excluded that the movements of the tower could be entirely ascribed to secondary settlements in confined conditions. Hence, the reasons of the continuous foundation problems were mainly attributed to a slow lateral plastic flow under high stress gradients within the soft silty clay layer, squeezed between the pile ends and the underlying sand, and to the possible progressive decay of timber piles.

3.3 Monitoring system

After the end of the diagnostic investigations, a comprehensive monitoring system was installed to analyze the deformation behavior and the structural conditions of the bell tower and the adjacent portion of the Basilica during all the phases of the strengthening interventions, consisting of:

– crack-gauges and long-base extensometers installed on the main cracks of the masonry walls;
– strain-gauges to measure the deformation of the steel cable installed in the bell-tower;
– thermal gauges to measure the temperature of the internal and external air, as well as inside the masonry at different distances from the outer wall;
– geotechnical instrumentation, including electrical piezometers, multibase extensometers and biaxial inclinometers;
– direct pendulum equipped with automatic telecoordinometer, for the measurement of the absolute horizontal movements of the top of the tower.

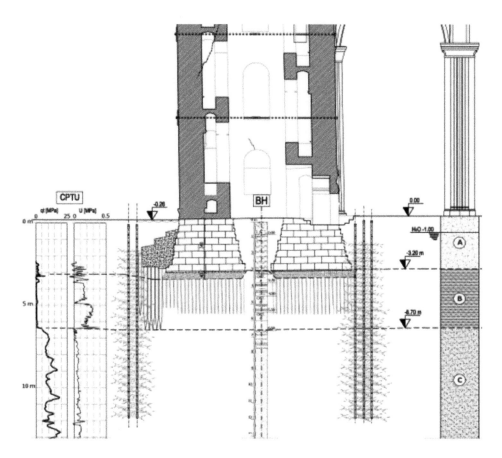

Figure 10. Schematic section of the tower foundation and of the relevant subsoil (SE-NW direction in Fig. 9), with the in situ test logs, the strengthening intervention of 1904 (left) and the TAMs of the new fracture grouting intervention (Gottardi et al., 2013).

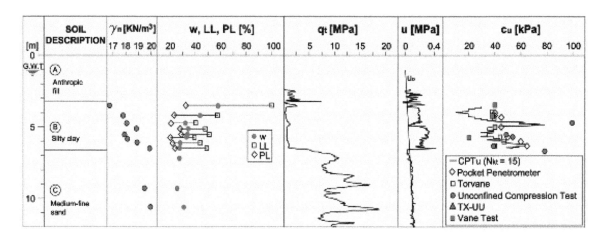

Figure 11. Upper soil profile with the relevant main geotechnical characteristics, as deduced from laboratory and in situ tests.

All the instruments were connected to an automatic data acquisition system that enabled to follow in real time the effect of the works on the structures, thus enabling to introduce possible suitable modifications to the intervention design. In addition, it was considered of vital importance to measure the vertical movements of the bell tower and of the adjacent portion of the Basilica. A high-precision and accurate manual leveling system with several measuring points was thus installed and periodical surveys were carried out and intensified during the most significant phases of the works.

4 THE INTERVENTION

The slow, but continuous, constant rate, differential settlement of the bell tower has soon become cause of major concern for the present and future stability, not only of the bell tower but, above all, of the structurally connected Basilica. From the results of the diagnostic investigations and the following numerical simulations (Lionello *et al.*, 2004), it clearly emerged that the interacting structures of the bell tower and the Basilica cannot bear further differential settlements without serious consequences.

It followed the need of a strengthening intervention, at the level of the foundations, aimed at reducing the differential settlements of the bell tower. A rather innovative intervention of careful soil fracturing, also known as fracture grouting, was eventually designed, in order to improve the mechanical characteristics of the soft silty clay. Once the aim of improving the stability of the soil-foundation system had been achieved, a new solution was required to reduce the damaging interaction between the masonry structures, activated by the foundation settlements. A structural joint between the bell tower and the Basilica was finally executed in order to improve the system deformability. In order to guarantee the safety of the whole Basilica and bell tower, a gradual strengthening intervention was designed, with a strict and constant control during the execution of the works, a rather typical and well implemented example of the so-called "Observational Method".

4.1 *Phase 1: provisional strengthening intervention*

A preliminary intervention was required by the concern of deformations induced on the column of the Basilica by the thrust of the bell tower. A provisional intervention was then carried out in order to increase the safety level of this specific and most delicate component of the Basilica. A steel cable was thus positioned connecting the stone ashlars just above the capital of the column (the one closest to the bell tower) to the bell tower structure at a height of 14.40 m), aimed at supporting part of the horizontal thrust acting on the column. Two strain-gauges were installed on the steel cable and the relevant tension constantly monitored during the whole intervention period.

4.2 *Phase 2: soil strengthening*

The principles on which the soil strengthening intervention was based were (Lionello, 2008):

- making compatible the remaining settlements of the complex Basilica and bell tower;
- preserving as much as possible the original foundation structure and the current stress distribution;
- avoiding a rigid foundation system for the bell tower;
- enabling a flexible and modular intervention, in constant agreement with the outcome of an extensive real time monitoring.

An intervention of fracture grouting was finally selected. This technique consists of installing special injection pipes (tubes à manchettes, i.e. TAMs, Fig. 12a) in the foundation soil, fitted with equally spaced valves at different depths. Each valve can be selectively injected by means of a double packer device (Fig. 12b). The careful and slow rate injection of suitable cement and bentonite mixtures can be repeated at successive stages, to obtain progressive increments of mechanical characteristics. The final outcome should be a reinforced soil, made up of the original material and an indented web of thin layers of injected grout (Fig. 12c). In order to evaluate the feasibility of the soil fracturing intervention (Mori & Tamura, 1987; Panah & Yanagisawa, 1989; Raabe & Esters, 1990; Andersen *et al.*, 1994; Alfaro & Wong, 2001; Soga *et al.*, 2005) and calibrate the design parameters (injection pressures, injection rate, grout mixture, etc.), a full-scale test site with geotechnical monitoring devices (piezometers and multibase extensometers) was carried out on the northern corner of the bell tower, inside the Basilica.

The test gave the expected results and the soil fracturing intervention was then carried out by means of eighty-eight 12 m-deep sleeve steel pipes (so-called "tubes à manchettes", i.e. TAMs) installed all around the perimeter of the bell tower (Fig. 13). Such special injection pipes had equally spaced (0.5 m) valves, selectively injectable by means of the double packer device. They were aligned along two rows, according to local geometric constraints, except for the west side where a third row was subsequently added. The inner row was placed as close as possible to the foundation side; the other two rows were spaced of about 50 cm

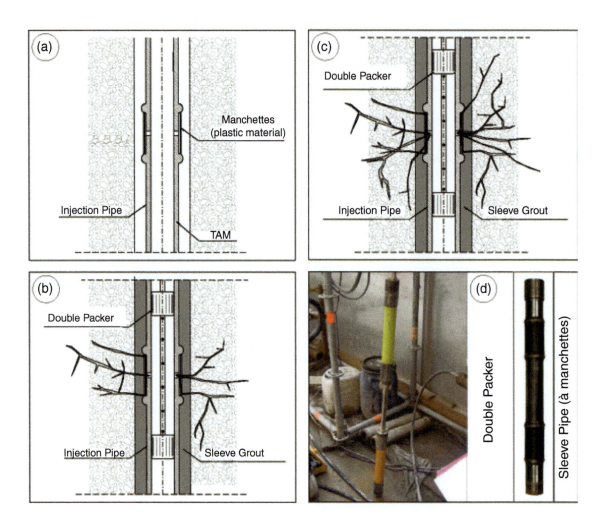

Figure 12. Soil fracturing via multiple injections: (a) initial state: injection pipe installed; (b) first injection: fractures predominantly in one direction; (c) continued injection: fractures in various directions; (d) double packer and sleeve pipe (TAM).

(Fig. 13). A cement-bentonite grout was injected from each valve in three main separate stages (cycles). A fourth cycle was designed to enhance the intervention on the clay layer only and for the 55 pipes that had registered a lower injection pressure in the previous stages. The selected grout was made up of water, cement, bentonite and calcareous fill, with a water/cement ratio of about 1.5 and a cement/bentonite ratio of about 14, thus producing a very low viscosity grout (1.5×10^{-2} Pa × s). During the intervention the flow rate was kept low and fixed at about 6 l/min.

The resulting valve opening pressure and the following steady-state injection pressure were always recorded. In the final cycle, a flow rate of about half than the previous cycles was used. Relevant injected volumes are shown in Figure 14. Constant grout volumes of 20 l/valve were injected each cycle in clay, whereas variable volumes between 14 and 20 l/valve in sand. In the 1st cycle all valves could be opened. In the 2nd cycle a few valves could not be opened in sand. In the 3rd and 4th cycles more restrictive criteria were adopted: a prefixed limit to the opening pressure of 500 kPa in clay and of 2500 kPa in sand, where therefore many more valves could not be opened. Figure 14 shows the total volumes injected in each cycle from each valve depth, divided by the total number of existing pipes (i.e. 88). Notice that the total amount of grout injected in clay is clearly greater than in sand. At the end of the intervention, a total of about 100 m³ of grout was injected. Figure 15 reports the average steady-state injection pressures (P_i) vs. depth for each cycle, as measured at the manometer. P_i is clearly greater in sand and tends to increase almost linearly with depth. Injection pressures vary between about 200–500 kPa in the first cycle to about 400–1200 kPa and more in the final cycle. In particular, the low

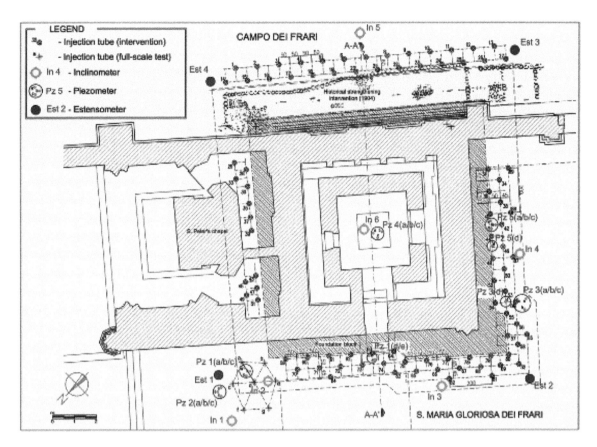

Figure 13. Plan of the Frari bell tower with TAMs layout and location of geotechnical monitoring devices (Lionello, 2008).

injection pressures (P_i) recorded in the third cycle between 2.00 and 3.00 m depth are not significant since they result from only 5 pipes.

In addition, P_i recorded in the fourth cycle, resulting from a reduced and selected number of pipes (55), turned out to be lower than that in the third cycle; this could be also due to the lower injection flow rate used in the final cycle, as already noticed in previous experiments. However, the clearly noticeable general increase of pressures with injection cycles reflects a corresponding increase of the soil minor principal stress during the intervention, as intended (Marchi et al., 2014).

4.3 Phase 3: structural joint

Once completed the intervention on the foundation soil and whilst continuing to monitor the bell tower behavior, the relevant structural interaction with the most damaged arches, vaults and columns of the Basilica had to be tackled.

Such interaction implied in fact:

– a substantial compression stress increment on the column, much greater than the contribution of the dead weight of walls, vaults and roofs;
– the formation of a large flexural stress on the column itself, due to the eccentricity of the vertical load and to the horizontal thrust component;
– the corresponding reduction of the vertical stress along the bell tower walls.

According to a suitably conceived approach for the restoration and preservation intervention, it was therefore opted, instead of opposing against the current forces acting on the structures, to reduce their stress distribution (Lionello, 2011). About two years after the end of strengthening works on the foundation soil, it was decided to insert a gap between the bell tower and the adjacent church by creating a structural joint that would allow the

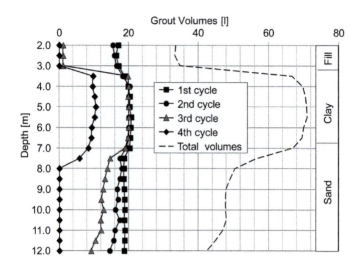

Figure 14. Average (for the overall 88 pipes) volumes injected from each valve, at selected depths, for the 1st, 2nd, 3rd and 4th cycle (Marchi *et al.*, 2014).

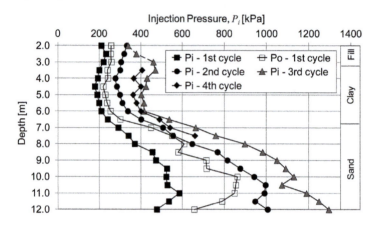

Figure 15. Average injection pressures vs. depth, for each cycle, as measured at the manometer (Marchi *et al.*, 2014).

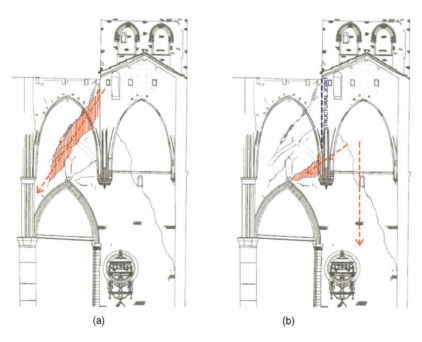

Figure 16. Direction of the thrust between the bell tower and the basilica: a) before the structural joint execution, b) after the structural joint execution.

relative movements and prevent the mutual stress transmission. It was assumed that the compression stresses flowing above the arch and transmitting substantial forces to the column of the Basilica (Fig. 16a) could be thus reduced.

The structural joint was created at the beginning of June 2008 in the position shown in Figure 13b and, in order to preserve the Basilica architecture, it was carried out only above the vaults where, on the other hand, mechanical interaction was greater. During the intervention, it could be confirmed that the existing structural link had been executed only after the construction, probably together with the early 20th century interventions, when it was wrongly pursued the approach of connecting the bell tower to the Basilica structures. The effect of such intervention was a marked change of direction of the thrust applied by the bell tower to the adjacent column, as shown in Figure 13b and predicted by numerical models. The execution of the structural joint was very slow and lasted about 6 months. During this period a detailed analysis of the information obtained by the monitoring system enabled to carry out the different steps of the intervention with a continuous check of the structural response, thus avoiding to induce damages to the bell tower and to the supporting structures of the Basilica.

5 INVESTIGATIONS AND MONITORING DURING AND AFTER THE INTERVENTION

5.1 *Soil investigations and monitoring*

As reported in Section 3.3, the geotechnical instrumentation installed throughout the area of the intervention enabled to carefully check the effects of the soil fracturing. As an example, in Figure 17 the records of most electric piezometers installed within the silty clay layer are provided.

The pore pressure peaks induced by adjacent grout injections are clearly visible as well as the subsequent relatively quick consolidation rate, which enabled to drive the operations under constant control and safety conditions. As designed, the injections led to soil fracturing: evident and diffuse cement lenses in the soil were found in undisturbed continuous core sampling carried out at the end of the intervention. Specific additional in situ geotechnical investigations (piezocone and dilatometer tests) were planned and carried out to assess its effect on the mechanical properties of soil. The global effect of the injections could be assessed by comparing the results of in situ tests performed at different stages of the intervention. The soil strength

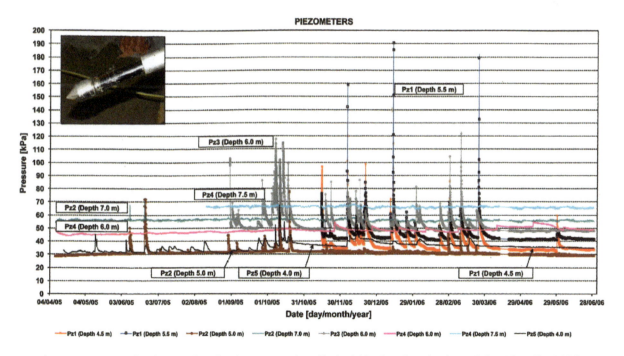

Figure 17. Records of most electric piezometers installed within the silty clay layer (after Lionello, 2008).

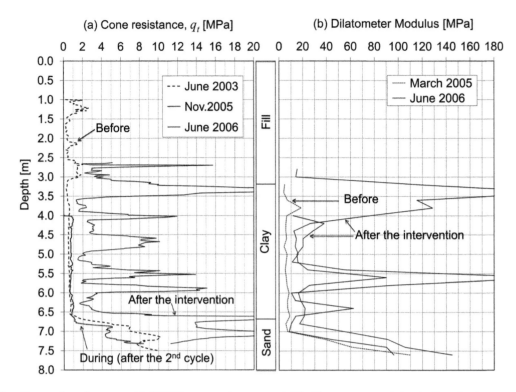

Figure 18. Comparison between (a) CPTU cone resistance and (b) dilatometer modulus before the intervention, after the second injection cycle and at the end of the injections (after Lionello, 2008).

improvement is clearly noticeable from all the CPTU tests carried out before the intervention, after the second injection cycle, and at the end of the injections (Fig. 18a) (respectively CPTu"n"/A, /D and /E in Fig. 9). Peaks of the tip resistance are due to the diffuse presence of grout lenses, although the relevant average increase among them was more effective. Figure 18(b) shows analogous data related to the dilatometer modulus. The relevant stiffness increase was also confirmed by laboratory oedometer tests, which provided a modulus increase of approximately 100% at 50 kPa and approximately 20% at 400 kPa (Lionello, 2008). However, more comprehensive information for the analysis of the structure behavior during the soil fracturing works come from the direct pendulum records and precision leveling surveys.

In Figure 19, the movement trend along two orthogonal components from direct pendulum are shown. The period of observation is from December 2003 (date of installation of the monitoring system) to the end of December 2016. During the soil fracturing intervention (from April 2005 to March 2006, grey shadow), the monitoring system was also useful to define the rate of the intervention phases as well as to support decisions on the parameters of the grouting procedures (injection pressure, flow rate, etc.). A significant movement of the bell tower during this intervention phase was observed, the component in x direction (toward the apse) being about 9.0 mm and in y direction (toward the Basilica) about 5.0 mm. After the end of the soil fracturing, the rate of the bell tower movements showed a quick reduction, reaching a lower value than that observed before the intervention. Movements of cracks in the stone arch which connects the bell tower to the Basilica also showed a marked increase during soil fracturing and a rapid decrease after.

In Figure 20 the settlement trend of several benchmarks from precision leveling is provided. A more pronounced vertical movement of the bell tower west and south corners (points 5 and 6) is clearly observed during the soil fracturing intervention (time interval highlighted in grey), but all measuring points display a similar rate increase during the intervention and a significant reduction immediately after, also with respect to original rates before, consistently with design expectations. Such observation is confirmed by the differential settlement trend (Fig. 20) along main alignments and, above all, between the critical column of the Basilica and the adjacent bell tower corner (points 1 and 5): 1.01 mm/year before, 5.04 mm/year during and 0.27 mm/year after the intervention (before the joint execution).

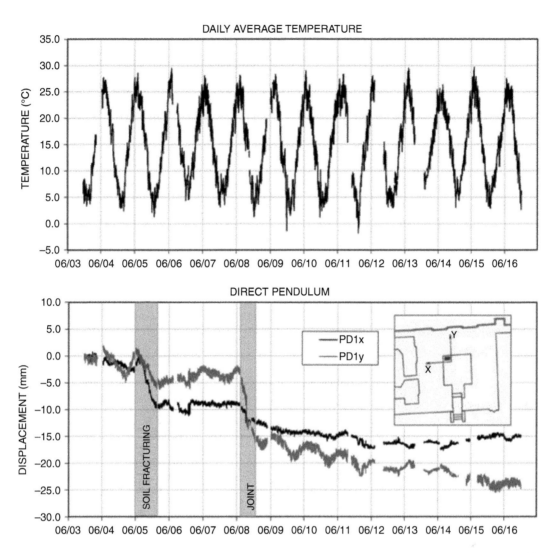

Figure 19. Complete history of the displacement components measured by the direct pendulum with the indication of the intervention phases: soil fracturing and structural joint execution (to 27/12/2016).

The settlement values are also in very good agreement with the measures obtained by the direct pendulum in the period from December 2003 to March 2007 (Fig. 21). Along the alignment of points 3 and 4, the differential settlements at the base of the bell tower (2.16 mm) multiplied for the ratio between the height and the base of the tower (4.70) turns out to be 10.15 mm, very close to the x component measured by the direct pendulum (10.00 mm). Along the alignment of points 5 and 10, the differential settlements at the base of the bell-tower (0.95 mm), multiplied for the same ratio, is equal to 4.46 mm, which is again very close to the y component measured by the direct pendulum (4.00 mm).

5.2 Structure investigations and monitoring

In order to follow with special care the deformation behavior during the execution of the structural joint, new crack-gauges as well as new long-base extensometers were installed. Furthermore, in order to check the thrust modification between the bell tower and the Basilica, special flat-jacks were installed in the positions indicated in Figure 22: on the right the stress values measured before the execution of the structural joint, while on the left the stress changes after the intervention are shown.

It can be observed a significant decrease of the state of stress in the upper part of the wall between the bell tower and the Basilica, which is a clear experimental confirmation of the assumptions on the thrust reduction applied by the bell tower.

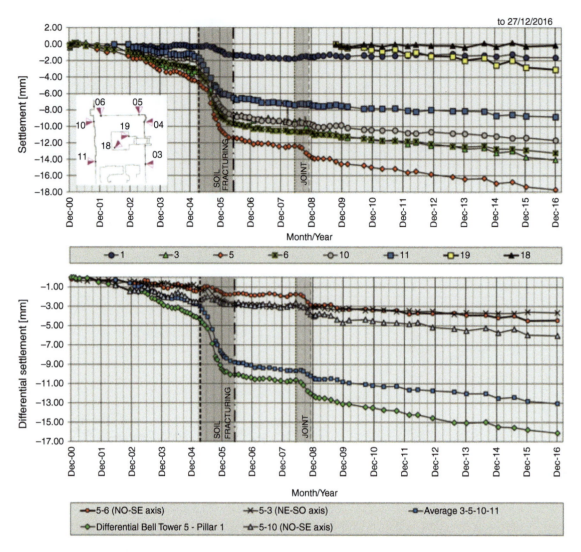

Figure 20. Settlements (top) and differential settlements (bottom) of several benchmarks from precision leveling (courtesy of FOART srl, Parma), to 27/12/2016.

In particular, the positive effects of the structural joint can be summarised as follows:

1. a substantial reduction of the load acting on the column of the Basilica, estimated as about 800 kN (−17%);
2. a corresponding increase at the base of the bell tower;
3. a restoration of the original conditions when the two structures were statically more independent.

An interesting indication of the bell tower response to such structural modification can come again from the measurements of the direct pendulum (Fig. 19, second shadowed time interval). It can be observed a more evident effect along the y direction, i.e. consistently with the newly introduced lack of support. It can be also helpful to analyse the movements of the masonry structures along the existing cracks together with the displacements provided by the direct pendulum and the settlements coming from precise levelling, especially when comparing the situation before (Fig. 23) and after (Figs. 24 and 25) the structural joint intervention.

In the time interval – about two years – between the soil fracturing and the structural joint, the tower has settled less than one millimeter (see surveying records at the top left of Fig. 23), with a modest inclination toward the external side. After the end of the structural joint execution and the complete removal of the temporary steel cables and of the timber prop system, the deformation behaviour has been constantly observed. The displacement components of the bell tower, measured by the direct pendulum, are more substantial and

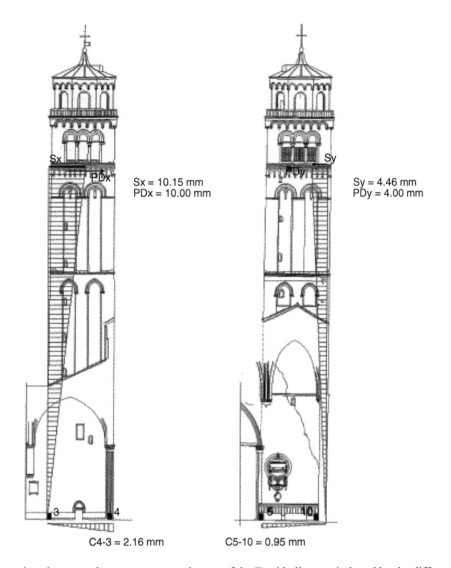

Figure 21. Comparison between the movements at the top of the Frari bell tower induced by the differential settlements of the foundations and those measured by the direct pendulum, in the period from December 2003 to March 2007 (Rossi & Rossi, 2012).

in the opposite direction, toward the Basilica. They now appear to be only partly related to the differential settlements of the foundation, the remaining part being due to deformation processes of the bell tower masonry.

Despite such substantial modification of the load distribution, however, which has inevitably produced a new increase (up to about 0.6 mm/year from January 2009 to November 2012, Fig. 20) of the differential settlement between the bell tower corner and the column (which incidentally has even showed a temporary upward relative movement, benchmark 1 in Fig. 20), all the monitoring devices (Fig. 19, 20, 23, 24 and 25) show a consistent trend. The average – and still rather uniform – settlement of the bell tower has proved to be about 0.35 mm/year, in the 4 years after the joint execution, from January 2009 to November 2012 (Gottardi et al., 2015), and about 0.3 mm/year, in the last 5 years, from February 2012 to December 2016.

When compared to the rate of more than 1 mm/year before the soil fracturing intervention, it appears that its overall effectiveness can be confirmed.

It is finally extremely interesting to compare the average foundation settlement during the 7 years after the joint execution with the readings of two adjacent benchmarks (18 and 19, Fig. 20, and in detail, Fig. 26) located inside the bell tower, one at the foundation level (+1.00) (benchmark 19) and the other – subsequently installed – anchored in the soil, between 5.5 and 6.00 m depth, just below the tip of timber piles (benchmark

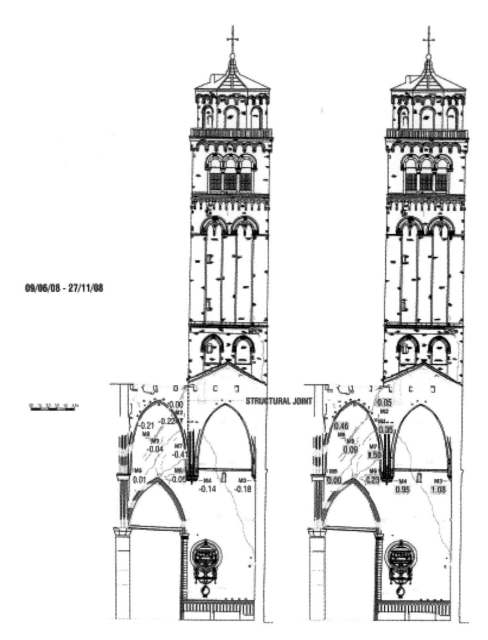

Figure 22. Measurements of the state of stress before the structural joint execution (on the right) and variations of the state of stress induced by the structural joint execution (on the left) (Rossi & Rossi, 2012).

18). Despite some scatter of the readings (see Fig. 20), due to the position of the benchmarks inside the bell tower, such comparison suggests that nearly all continuous soil straining (almost 100% of the overall foundation settlement) occurs between the ground level and the end of piles. This evidence would suggest that the progressive and somehow expected decay of timber piles – due to the very poor quality of wood extracted from relevant inclined boreholes, despite its preservation in constantly full anaerobic conditions – can count for most residual movements of the bell tower.

Thanks to the current reduced static interaction, however, the bell tower can accommodate further movements without inducing excessive stresses on the adjacent Basilica. The separation of the two structures has proved to be especially positive also in relation to seismic actions, if after the shakes of the Emilia earthquake in May 2012 no specific consequences have been recorded in the isolated structure, whilst new cracks have appeared in the nearby St. Peter Chapel, still interested by structural links.

Careful monitoring of settlement rates will continue in order to keep under control possible future trends and to fully understand the ongoing mechanisms.

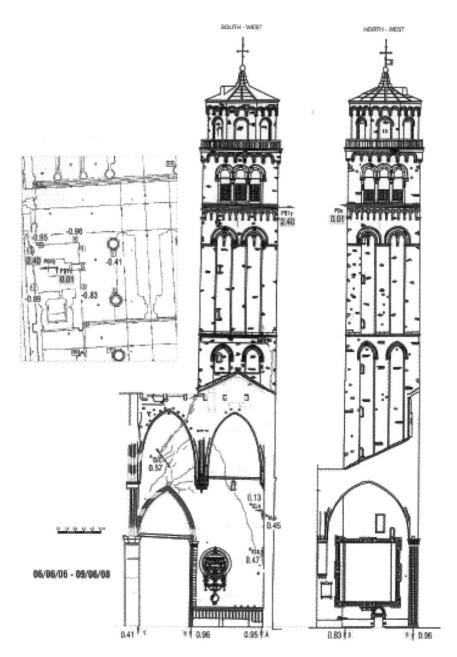

Figure 23. Analysis of the deformations measured during a period of two years, between the soil fracturing intervention and the structural joint execution.

6 CONCLUSIONS

Venice and its more than 80 bell towers are a vulnerable precious piece of the Italian cultural and artistic heritage. Neglected until the collapse of San Marco bell tower in 1902, they have now become central in monitoring and maintenance programs carried out by local public authorities. The full respect of their historic, architectural and structural features is the basic principle of any possible intervention.

An emblematic example of such an approach is represented by the multiphase intervention carried out on the bell tower of the Frari complex, the second tallest in Venice, which since its construction has been affected by slow but constant differential settlements between the tower itself and the adjacent masonry structures of the Basilica. At the end of the last century, a growing concern for the stability of the structures involved provided the impetus for the development of modern remedial measures, first in foundation and then on the elevation

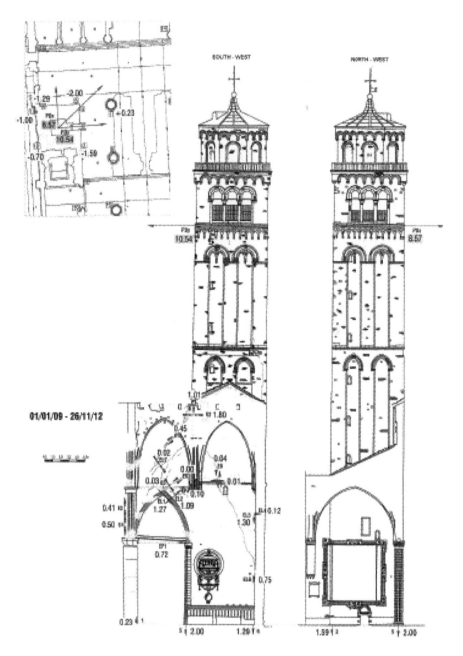

Figure 24. Analysis of the deformations measured during a period of about four years, after the structural joint execution.

structure. A ground modification intervention by soil fracturing was then carried out in order to improve the mechanical characteristics of the clayey layer underlying the tower. Once the aim of improving the stability of the soil-foundation system had been achieved, a new solution to reduce the damaging interaction between the masonry structures, activated by the foundation settlements, was required. A structural joint between the bell tower and the Basilica was finally executed in order to improve the system deformability. On the other hand, in the design of the intervention, any possible action producing a stiffness increase of the elevated structures, thus altering the overall structural behavior, has been carefully prevented.

As regards the foundations, innovative criteria and methodologies were used, so far mainly devoted to rehabilitation and strengthening works on the upper structures. The preservation of the foundation material has been fully guaranteed, without any direct interventions on the relevant structures, by the insertion into the surrounding soil of injection pipes for soil fracturing. Through the articulated and extensive real-time monitoring system, purposely implemented, it was possible to focus on the intervention areas, calibrating and

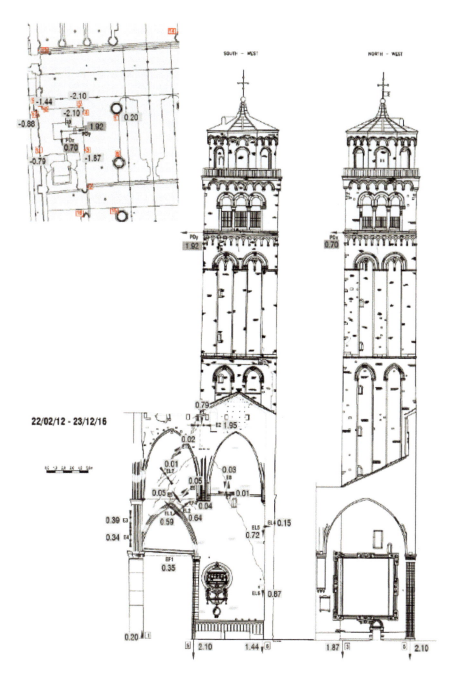

Figure 25. Analysis of the deformations and displacements measured in the last five years (February 2012–December 2016).

minimizing the amount of injected grout. The technology used has proved to be especially flexible and it will enable further injection cycles in the future, if needed.

The overall monitoring carried out for about 15 years has confirmed the good compatibility of the strengthening intervention on the bell tower with the mechanical characteristics of the adjacent Basilica. The tower is currently moving more vertically and with a settlement rate substantially reduced, from almost 1 mm/year before the intervention works to 0.3 mm/year after them. In addition, the current reduced static interaction – due to execution of the joint in the elevation structure – enables the bell tower to accommodate further movements without inducing excessive stresses on the adjacent Basilica. The assestimeter installed at 5.5 m depth, just below the tip of timber piles, clearly shows that most of the present residual movements of the bell tower can be ascribed to the progressive decay of foundation piles, which should be therefore further and carefully investigated.

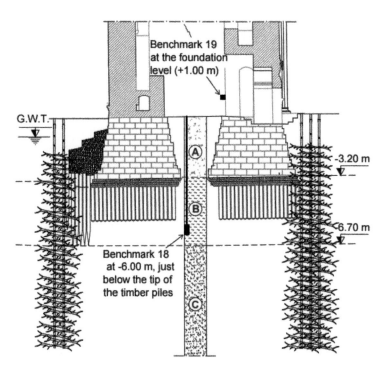

Figure 26. Schematic section of the foundation with the position of the deep assestimeter (benchmark 18) and of benchmark 19.

REFERENCES

Alfaro, M.C. & Wong, R.C.K. (2001). Laboratory studies on fracturing of low permeability soils. *Canadian Geotechnical Journal*, 38, 303–315.

Andersen, K.H., Rawlings, C.G., Lunne, T.A. & By, T.H. (1994). Estimation of hydraulics fracture pressure in clay. *Canadian Geotechnical Journal*, 31, 817–828.

Gottardi G., Cavallari L. & Marchi M. (2009). Soil fracturing of soft silty clays for the reinforcement of a bell tower foundation. *2nd International Workshop on Geotechnics of Soft Soils – Focus On Ground Improvement, University of Strathclyde, Glasgow, Scotland, 3–5 September 2008.* London: Taylor & Francis Group, pp. 31–41.

Gottardi, G., Lionello, A., Marchi, M. & Rossi, P.P. (2013). Preservation and monitoring of the Frari Bell Tower in Venice. *2nd International Symposium on Geotechnical Engineering for the Preservation of Monuments and Historic Sites (IS TC301). Napoli, Italy, 30–31 May 2013*, Taylor and Francis Group, London, UK, pp. 411–424.

Gottardi, G., Lionello, A., Marchi, M. & Rossi, P.P. (2015). Monitoring-driven design of a multiphase intervention for the preservation of the Frari bell tower in Venice. *Rivista Italiana di Geotecnica*, 49 (1), 45–64.

Levi, C. A. (1890). I campanili di Venezia, Notizie storiche. Ferdinando Ongania Ed., Venezia (in Italian).

Lionello, A., Cavaggioni, I., Marchi, G., Gottardi, G., Ragazzini, A., Modena, C., Casarin, F., Rossi, P.P. & Rossi, C. (2007). Monitoraggio e controllo del consolidamento della fondazione del campanile dei Frari a Venezia. *Previsione e controllo del comportamento delle opere; XXIII Convegno Nazionale di Geotecnica, Abano Terme, 16–18 Maggio 2007.* Bologna: Patron Ed., pp. 585–592 (In Italian).

Lionello, A., Cavaggioni, I., Rossi, P.P., Rossi, C., Modena, C., Casarin, F., Marchi, G., Gottardi, G. & Ragazzini A. (2004). Preliminary investigations and monitoring for the design of a strengthening intervention on the Frari Basilica, Venice. *Possibilities of numerical and experimental techniques, IV International Seminar on Structural Analysis of Historical Constructions (SAHC), Padova, 10–13 Novembre 2004.* Balkema, pp. 1323–1333.

Lionello, A. (ed.) (2008). Il campanile di Santa Maria Gloriosa dei Frari in Venezia – conoscenza, consolidamento e restauro. Milano: Electa (in Italian).

Lionello, A. (ed.) (2011). Tecniche costruttive, dissesti e consolidamenti dei campanili di Venezia. Venezia: Corbo e Fiore Publishers (in Italian).

Marchi, M., Gottardi, G. & Lionello A. (2006). Sulle fondazioni dei campanili di Venezia. Fondazioni Superficiali e Profonde, *V Convegno Nazionale dei Ricercatori di Ingegneria Geotecnica, Bari, 15–16 Settembre 2006.* Benevento: Hevelius Ed., Vol. 1, pp. 177–192 (in Italian).

Marchi, M., Gottardi, G., Ragazzini, A. & Marchi, G. (2012). On the sand response of fracture grouting on a bell tower foundation. *Int. Conf. on Ground Improvement and Ground Control (ICGI), 30 Oct.–2 Nov. 2012, University of Wollongong, Australia.* Vol. 2, pp. 1319–1325.

Marchi, M., Gottardi, G. & Soga, K. (2014). Fracturing pressure in clay. *Journal of Geotechnical and Geoenvironmental Engineering (ASCE)*, 140 (2), 04013008(1-9).

Marchi, M., Gottardi, G. & Soga, K. (2015). Closure to fracturing pressure in clay. *Journal of Geotechnical and Geoenvironmental Engineering (ASCE)*, Vol. 141 (5), pp. 07014045–07014045.

Mori, A. & Tamura, M. (1987). Hydrofracturing pressure of cohesive soils. *Soils and Foundations*, 27 (1), 14–22.

Panah, A.K. & Yanagisawa, E. (1989). Laboratory studies on hydraulic fracturing criteria in soil. *Soils and foundations*, 29 (4), 14–22.

Raabe, E.W. & Esters, K. (1990). Soil fracturing techniques for terminating settlements and restoring levels of buildings and structures. M.P. Moseley (ed.), *Ground Improvement*, Glasgow: Blackie A & P., pp. 175–192.

Rossi, P.P. & Rossi, C. (2012). Monitoring of two great venetian cathedrals: San Marco and Santa Maria Gloriosa dei Frari. *Proc. of the Int. workshop on Monitoring of Great Historical Structures. Florence, January 2012.*

Soga, K., Ng, M.Y.A., & Gafar, K. (2005). Soil fractures in grouting. *Proc. of the 11th International Congress on Computer Methods and Advances in Geomechanics, Turin, Italy*, pp. 397–406.

Zuccolo, G. (1975). Il restauro statico nell'architettura di Venezia. Istituto Veneto di Scienza Lettere ed Arti, Venezia (in Italian).

Geotechnics and Heritage: Historic Towers – Lancellotta, Flora & Viggiani
© 2018 Taylor & Francis Group, London, ISBN 978-1-138-03272-9

Carmine Bell Tower in Napoli: Prediction of soil–structure interaction under seismic actions

F. de Silva
Department of Civil, Architectural and Environmental Engineering, University of Napoli Federico II, Napoli, Italy

F. Ceroni
Department of Engineering, University of Napoli Parthenope, Napoli, Italy

S. Sica
Department of Engineering, University of Sannio, Benevento, Italy

F. Silvestri
Department of Civil, Architectural and Environmental Engineering, University of Napoli Federico II, Napoli, Italy

ABSTRACT: Evidence of damage induced by past and recent earthquakes in Italy testifies to the seismic vulnerability of masonry towers, especially when founded on soft soils. A significant case study is represented by the 68 m-high Carmine Bell Tower, located in the coastal area of Napoli. The structure, the foundation and the subsoil of the tower were thoroughly investigated through field and laboratory tests and dynamic monitoring. Nonlinear seismic response analyses of the soil–foundation–structure system were carried out on a full dynamic 3D model, for which the parameters were accurately calibrated from experimental data. Because the tower stands between a church and a friary, the effect of the lateral constraints exerted by the neighbouring structures was numerically investigated. The lateral constraint imposed along the tower was found to strongly affect the seismic response of the system, as it changes the position and the time of occurrence of the plastic zone that takes place in the structure or the foundations. Finally, the foundation safety was assessed by comparing the evolution of the base forces to the failure locus.

1 INTRODUCTION

The widespread damage caused by the recent strong-motion earthquakes in Abruzzo (2009) and Emilia (2012), as well as by the most recent seismic sequence in Central Italy (2016), testifies to the seismic vulnerability of the Italian environment and its building heritage, and the need for sustainable mitigatory countermeasures to ensure their conservation and fruition.

The aforementioned seismic events significantly hit outstanding historical centres, most of them dating back to the Middle Ages; some examples are L'Aquila in 2009 (Monaco *et al.*, 2012), Mirandola in 2012 (Fioravante et al., 2013), and Amatrice in 2016 (Stewart *et al.*, 2016). The damage observed on the masonry buildings particularly highlighted the vulnerability of the monumental heritage to earthquakes. Many churches and bell towers were seriously damaged, with great anguish for local populations, because these iconic structures represent the historical, social and religious identities of the communities living around them. Among the examples, the case of Mirandola tower, destroyed by the M_W 6.1 Emilia earthquake (20 May 2012), is emblematic because soil–structure resonance was recognised as triggering the damage of the tower, highlighting the role of the soil–foundation–structure (SFS) interaction. Thus, the assessment and improvement of the seismic safety of a historical tower requires an accurate analysis of the dynamic response of the SFS system, which is still overlooked by much current practice.

Statistics that we have collected on the Italian monumental towers show that the seismic hazard affecting them is typically high (for most of them, reference accelerations greater than $0.1g$ are expected with a 475-year return period), and that the subsoil can often be very deformable (de Silva *et al.*, 2015). As is well known, in such conditions the seismic action transmitted by the foundation to the structure differs from the free-field motion (kinematic interaction), while the absolute structural displacement is increased due to the subsoil compliance (inertial interaction). With respect to a fixed-base structure, the fundamental period of the SFS system increases and part of the seismic energy is radiated into the soil, usually reducing the flexural displacement and the structural demand.

The present chapter illustrates the numerical procedure developed to model the SFS interaction of the Carmine Bell Tower in Napoli, originally destroyed by a strong earthquake in 1456 and thereafter reconstructed. With its 68 m height, it is the highest historical construction of the city, with an outstanding iconic significance. After a summary of the investigations on the structure and the subsoil (Section 2), the full dynamic model created with the Finite Difference Method (FDM) through the FLAC3D™ software is described (Section 3) and validated through a dynamic identification procedure (Section 4), with the aim of individuating the type of lateral constraint exerted on the tower by the surrounding buildings, in order to better represent the experimental behaviour detected by the dynamic structural monitoring. Seismic, nonlinear SFS interaction analyses were carried out by considering an input motion representing the maximum historical earthquake that occurred in 1456, with a particular emphasis on the role of the base deformability and the lateral constraints imposed on the tower (Section 5). Finally, the foundation safety was assessed by comparing the evolution of the base forces to the failure locus (Section 6).

2   HISTORY AND INVESTIGATIONS ON THE CARMINE BELL TOWER IN NAPOLI

The case study analysed here is that of the Carmine Bell Tower, a very slender tower on deformable soil in the eastern seashore of Napoli. Together with a church, a friary and a cloister (see Figures 1a and 1b), the bell tower comprises an important monumental complex, holding a significant historical and symbolic value for the city.

The complex is located in the market area of the ancient city, close to the main gate of the original Aragon walls. According to ancient chronicles (Celano, 1856; Galante, 1873), in the 7th century Father Ludovico da Casoria established a hospice near the sea, outside the walls of the city. A small chapel was built next to it, representing the initial core of the future Santa Maria del Carmine church. In the 12th century, Carmelite friars brought an image of the Virgin and Child to Napoli, which was hosted in the chapel and venerated as "Madonna Bruna" because of her dark face. Thanks to the alms received for the Virgin, in the 13th century the friars built a larger church and a convent. The bell tower was probably built in the course of these renovation works, even though the first reference to its presence dates back only to the 15th century (Moscarella, 1589).

In 1456, a destructive earthquake with a moment magnitude, M_W, estimated as high as 7.2 induced a macroseismic intensity $I_{MCS} = 8$ at Napoli. Figure 1c shows the distribution of I_{MCS}, derived by Rovida et al. (2011): the affected area extends for about 18,000 km^2, across Central-Southern Italy and from the Adriatic to the Tyrrhenian coasts. Although the epicentre was as far as 63 km north-west of Napoli (see Figure 1c), the existing bell tower was severely damaged and later reconstructed on the residual basement. The current bell tower has three stages with a quadrilateral cross section, an octagonal cell and a pyramidal spire (see Figures 2a and 2b). From the basement up to 40 m, that is, while the cross section is square, the faced masonry walls are made of yellow tuff, while clay bricks were adopted for the octagonal cell and the dome (Ceroni et al., 2009).

The "Madonna Bruna del Monte Carmelo" is still celebrated every year, on 16th July, with a characteristic popular feast based on sacred rites and secular traditions. The votive image is brought to the square for a Mass, after which fireworks are set off from the bell tower to simulate its burning. Some pictures of the celebrations are shown in Figure 3.

Geophysical surveys – Electrical Resistivity Tomography (ERT) and Ground-Penetrating Radar (GPR) – together with vertical and inclined boreholes, have recently been conducted to investigate the foundation, as reported in detail by de Silva et al. (2015) and Evangelista et al. (2016). The foundation of the bell tower coincides with the E–W main walls, which broaden 2 m below ground level (see Figure 2d). The foundation widens out about 30 cm in the middle part and 50 cm in the corners that is, approximately one and two 'spans' according to the Aragonese measurement system (1 span ≈ 0.26 m; Afan de Rivera, 1840). Even though the bell tower is located between other constructions, historical investigations (Ceroni et al., 2009) suggest that any structural connection be excluded, so that both the interactions with the church on the north side of the tower (up to an elevation of 19 m) and with the friary on the south side (up to 16 m) can be assumed as providing only contact constraints.

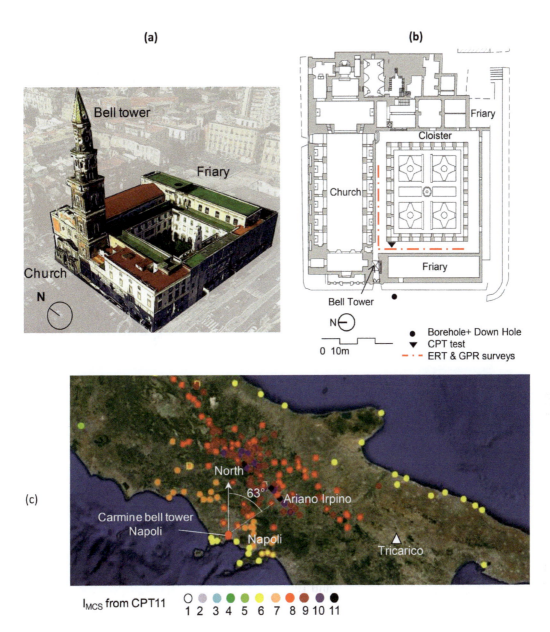

Figure 1. Aerial view (a) and plan with the location of the surveys (b) of the Carmine monumental complex; (c) map of the epicentre and intensity distribution of the 1456 earthquake (satellite images (a) and (c) courtesy of Google).

Dynamic structural tests were carried out, with three days of continuous monitoring of the tower under environmental actions (traffic, bell-ringing, wind, and human activities) and localised impulses induced by an instrumented hammer (Ceroni et al., 2009). Through the basic instruments of Operational Modal Analysis, the two main frequencies of the tower and the corresponding modal shapes were reliably identified. The associated modes were both translational and uncoupled: the first modal shape was parallel to the north–south direction, with a natural frequency of 0.68 Hz, while the second was parallel to the east–west direction, with a frequency of 0.76 Hz. Higher-order bending modes were experimentally identified at frequencies of 2.28 Hz and 2.35 Hz in the N–S and E–W directions, respectively.

The subsoil investigations, described in detail by de Silva et al. (2015), included a deep borehole drilled to 59 m in depth (shown by a small black circle in Figure 1b), very close to the external access to the bell tower. The lithological sequence is shown in Figure 4a: Man-made Ground (MG) down to 10 m in depth; thereafter, Marine Sand (MS) down to 31 m, interbedded with Pyroclastic Soil (PS) constituted by volcanic ash lenses and pyroclastic silty sand ('pozzolana'). The deepest investigated deposit is a layer of slightly cemented Yellow Tuff (YT), followed by Green Tuff (GT) characterised by similar lithological properties.

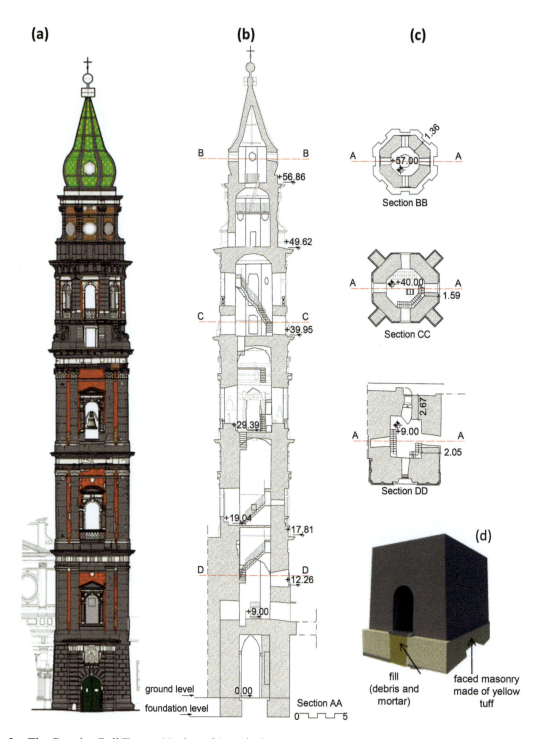

Figure 2. The Carmine Bell Tower: (a) view; (b) vertical cross section; (c) horizontal cross sections; (d) foundation.

The water table was intercepted at a depth of 2 m, exactly at the foundation level of the tower. A down-hole test was performed in the borehole to measure the compression (V_P) and shear (V_S) wave velocity down to 56 m, and the profiles are plotted in Figure 4b. Immediately below ground level, the high value of V_S (500 m/s) corresponds to the Aragonese walls. Thereafter, V_S increases with depth in the underlying man-made ground (MG), and then reduces to an almost constant value (about 300 m/s) in the upper layer of marine sand (MS). In the tuff formation the values of V_S gradually increase from 650 m/s at the top of the YT to 785 m/s in the GT.

To investigate the soil strength properties, a cone penetration test (CPT) was performed in the cloister down to a depth of 15 m (see Figures 1c and 4c). Consistent with the heterogeneity of the man-made ground

Figure 3. Santa Maria del Carmine celebrations: (a) Mass; (b) fireworks set off on the bell tower.

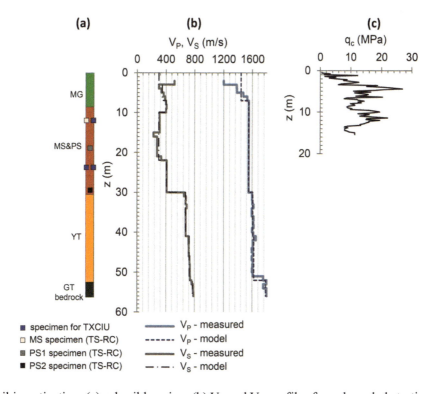

Figure 4. Subsoil investigation: (a) subsoil layering; (b) V_P and V_S profiles from down-hole testing; (c) tip resistance profile from cone penetration test.

intercepted by the borehole, the tip resistance, q_c, is very variable (3–27 MPa) along the first 8 m, while minor oscillations are shown in the marine sand (mean $q_c = 15$ MPa) and through the pyroclastic soil (mean $q_c = 8$ MPa). Triaxial consolidated-undrained (TX-CIU) tests were performed on the three samples shown in Figure 4a, in order to define the shear strength of the lower MS and PS deposits. To characterise the nonlinear and dissipative soil behaviour, undrained Resonant Column (RC) and Torsional Shear (TS) tests were performed on three saturated specimens taken at 12 m depth in the marine sand (MS in Figure 4a), as well as at 16 m and 30 m depths in the pyroclastic soil (PS1 and PS2 in Figure 4a). The corresponding data

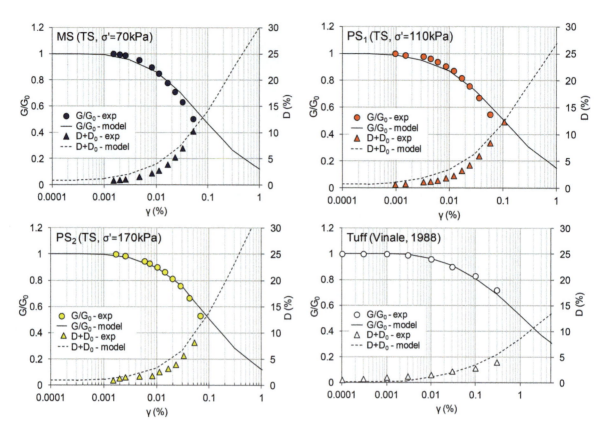

Figure 5. Variation with shear strain, γ, of normalised shear modulus, G/G_0, and damping ratio, D: experimental data and analytical curves obtained through a sigmoidal model.

points describing the variation with shear strain, γ, of the normalised shear modulus, G/G_0, and the damping ratio, D, resulting from these TS tests are plotted in the charts labelled with MS, PS$_1$ and PS$_2$ in Figure 5.

3 FULL DYNAMIC MODEL OF SOIL, FOUNDATION AND STRUCTURE

The three-dimensional model shown in Figure 6 was realised through the finite difference code FLAC3D 5.01 (ITASCA, 2015). The subsoil domain extends 50 m × 50 m in plan and 52 m in depth; it is composed of the layering shown in Figure 4a, by assuming the green tuff formation as an underlying elastic half-space. Site amplifications related to the subsoil morphology are not expected, due to the horizontal soil layering. The water table was set at 2 m below the ground surface, as detected on site.

The model reproduces the geometry of the tower structure in detail, including the barrel vault on the ground floor and the covering dome. The thickness of the walls varies with height, from 2.5 m to 1 m. The masonry walls were discretised through brick elements, with thickness decreasing according to height. The barrel vault at the first level and the dome were respectively simulated through a radial-cylindrical shape and a pyramid, both included in the FLAC3D library. The foundation was 2 m deep and 3.5 m wide, with an enlargement of 0.50 m, with respect to the ground floor, along each side of the main walls.

In order to avoid undesired wave reflections, 'quiet boundaries' (i.e. absorbing horizontal and vertical dashpots) were applied at the base of the layering, while 'free-field boundaries' (i.e. soil columns reproducing one-dimensional motion) were set along the lateral sides and at the corners. These boundary conditions minimise the size of the domain of analysis and reduce calculation time significantly.

Physical and mechanical properties of the soil and the superstructure were all derived from the field surveys and laboratory measurements, as summarised in the previous section and described in detail by Ceroni *et al.* (2009) and de Silva *et al.* (2015). The values adopted for the parameters are reported in Table 1 for the soil and Table 2 for the structure.

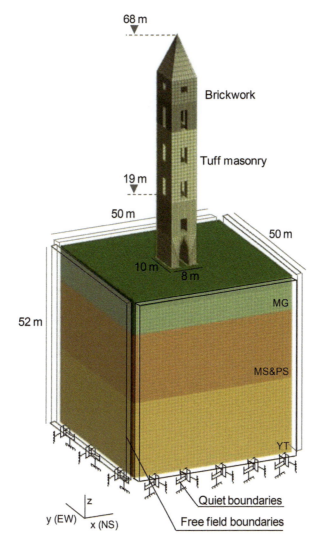

Figure 6. Global view of the 3D finite difference model.

The soil elasticity properties were computed from the V_P and V_S profiles derived from the down-hole test (Figure 4b). A Mohr–Coulomb strength criterion was adopted for the soil layers, with the friction angle, φ, derived by the CPT test for the man-made ground (MG) and calculated from the TX-CIU tests for the marine and pyroclastic sandy soils (MS and PS). For both of them, the dilation and the tensile strength were assumed null and a cohesion $c' = 2$ kPa was set to improve the stability of calculations at low values of the mean stress.

A hysteretic pre-failure behaviour was assumed for the soil layers, modelling the variation with the shear strain of the shear modulus and the damping through 'sigmoidal' functions. The parameters describing the shape of the functions were calibrated through best-fitting of the stress–strain loops measured in the torsional shear tests. In Figure 5, the resulting curves are compared to the equivalent parameters obtained from the torsional shear tests. The behaviour of the most superficial MS sample was assumed as representative for the man-made ground, while the curves obtained for PS_1 and PS_2 were, respectively, associated with the upper and lower layers of pyroclastic soil and marine sand. The yellow tuff was modelled as a hysteretic material too, calibrating the parameters of the sigmoidal function on the resonant column data reported by Vinale (1988). The fair agreement between the experimental data and the response of the constitutive model is demonstrated in Figure 5.

Following the well-known Rayleigh (1945) approach, a small-strain damping ratio, D_0, variable with frequency, was assigned to each soil layer. The damping–frequency relationship was calibrated to yield a minimum value of D_0 equal to that measured in the RC tests, in correspondence with the fundamental frequency of the layered deposit.

Table 1. Soil mechanical properties.

Soil		Constitutive model	γ kN/m³	V_P m/s	V_S m/s	G_0 MPa	ν –	K_b MPa	D_0 %	φ °	c kPa	σ_t MPa
MG	Upper	Hysteretic	20.7	1440	300	187	0.45	1809	0.83	38	2	0
	Medium	Mohr–		1440	346	249	0.45	2406	0.83			
	Lower	Coulomb		1550	401	335	0.45	3237	0.83			
MS & PS	Upper	Hysteretic	20.60	1550	281	162	0.48	4000	0.65	36	2	0
	Lower	Mohr–Coulomb	16.40	1550	412	278	0.46	3385	0.93			
YT	Upper	Hysteretic	17.20	1608	677	789	0.39	3322	0.15	–	–	–
	Lower			1622	730	917	0.37	3221	0.15			
GT	(Bedrock)	Elastic	19.12	–	756	1114	–	–	–	–	–	–

Table 2. Structural mechanical properties.

Material		Constitutive model	γ kN/m³	G_0 MPa	ν –	K_b MPa	ξ %	φ °	c MPa	σ_{cy} MPa	σ_t MPa
Masonry	Brick	Double-yield	16.3	522	0.15	571	5	34	0.64	2.4	0.24
	Tuff	Double-yield	11.2	391	0.15	429	5	33	0.38	1.4	0.21
Foundation	Tuff	Double-yield	11.2	333	0.15	364	5	33	0.38	1.4	0.21

The values of the Young's modulus of the tuff and brick masonry, E = 900 MPa and E = 1200 MPa respectively, were assumed equal to those inferred from the structural dynamic identification, assuming the Poisson's coefficient $\nu = 0.15$ (Ceroni et al., 2009). More details of the geometry, the boundary conditions and the properties of the elastic field of the model are reported by de Silva (2016) and de Silva et al. (2017). A constant structural damping $\xi = 5\%$ was assumed for both the masonry and the foundation (see Table 2).

The same elasto-plastic strain-softening constitutive models adopted by Ceroni et al. (2010) were assumed for the brick and tuff masonry. Such behaviour was simulated through the 'Double-yield' model, included in the FLAC3D library. The model consists of a Mohr–Coulomb shear failure locus, in which the tension is limited by a cut-off, σ_t, and a compressive failure locus is introduced through the limit mean stress, p_c. The friction angle, φ, was derived from the relationships reported by Augenti and Parisi (2010) for the tuff, and from the suggestions by Atkinson (1989) for the brick masonry, while the cohesion was back-calculated as follows:

$$c = \frac{\sigma_{cy}}{2\sqrt{\dfrac{1 + \sin\phi}{1 - \sin\phi}}} \qquad (1)$$

where σ_{cy} is the peak uniaxial compression strength. The values of σ_{cy} were conservatively set as 1.4 MPa for the tuff and 2.4 MPa for the brick masonry, according to the results of the double flat-jack tests performed on the bell tower by Ceroni et al. (2009).

According to the experimental tests on the tuff masonry by Augenti and Parisi (2010), the residual compression strength was assumed equal to 50% of the peak strength for the tuff and reduced to 40% for the brick masonry, while the maximum strain in compression was assumed equal to 1% for both materials. Such a limit corresponds to a value of the ductility factor $\mu = 6$, as found in the study by Augenti and Parisi (2010).

For lack of experimental results, the tensile peak strengths were conservatively fixed as equal to 10% (for the brick masonry) and 15% (for the tuff) of the compression strength. After the peak, a linear softening until zero-stress at strains of 0.04% (for the tuff) and 0.05% (for the brick masonry) was hypothesised, with the final maximum strain in tension equal to 0.1%. The uniaxial stress–strain target laws are shown in Figure 7.

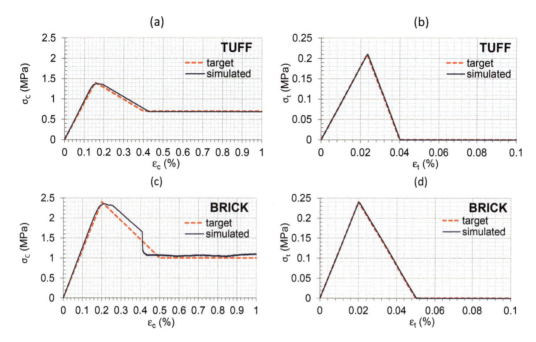

Figure 7. Target vs. simulated test behaviours of masonry: (a) tuff under uniaxial compression; (b) tuff under tension; (c) brick under uniaxial compression; (d) brick under tension.

The compressive bound of the constitutive model, p_c, was set equal to the mean stress during the unconfined uniaxial compression test, that is, $p_c = \sigma_{cy}/3$. To verify the constitutive models adopted, uniaxial tension and compression tests were simulated on a cylindrical sample. The brittle stress–strain behaviour simulated by the 'Double-yield' model resulted in good agreement with the target constitutive relationships, as shown in Figure 5.

Interface elements were applied to the underground lateral sides and at the base of the tower, where the foundation and the soil are in contact. The interface constitutive model was defined by a linear Mohr–Coulomb shear strength criterion. The values of friction angle, cohesion, dilation and tensile strength assigned to the interface elements were the same as the MG soil deposit (Table 1). Linear and nonlinear dynamic analyses, performed on the model with and without the interface, yielded the same results, confirming that the interface does not affect the model response, but is only a useful instrument to directly monitor the foundation behaviour.

4 NUMERICAL DYNAMIC IDENTIFICATION

The dynamic behaviour of the tower was simulated through the full model shown in Figure 6 by considering two different structural patterns:

- Free Tower, FT, a simple cantilever scheme;
- Restrained Tower, RT, with rotations in the vertical planes prevented between 0 and 19 m.

An original numerical procedure was developed to identify the frequency response of each pattern. Assuming linear elastic behaviour for all the materials, both patterns were excited by a random noise signal lasting 5 s, acting separately along the base nodes in the x and y directions. The structural response was predicted for 10 s, in order to record the free vibration behaviour in the last 5 s. At each elevation z of the structure, the flexural displacement, u_f, was computed as:

$$u_f = u_{ss} - (u_0 + u_\theta) \quad (2)$$

where u_{ss} is the total displacement, u_0 is the translation of the foundation, and u_θ is the displacement induced by the rigid rotation of the foundation, as shown in Figure 8a. The fast Fourier transforms of the flexural

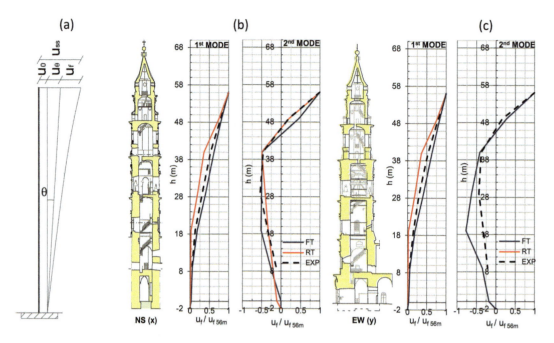

Figure 8. (a) Displacement components. Displacement plots for the experimental vs modelled free tower (FT) and restrained tower (RT) cases along: (b) *x*-axis; (c) *y*-axis.

displacements were computed and, through an Operational Modal Analysis (OMA), the vibration modes identified in correspondence of the peak amplitude values. In Table 3, the resulting numerical frequencies are listed for the FT and RT patterns, together with the experimental values reported by Ceroni *et al.* (2009). It is apparent that the first two experimental frequencies are significantly underestimated, that is, by about 50%, when using the free tower (FT) pattern. Instead, the better reproduction of the observed behaviour is obtained by introducing the structural restraints (RT), with differences as low as −17% and −5% for the *x* and *y* directions, respectively. For the second modes along *x* and *y* directions, the free tower pattern underestimates the experimental values by about 25%, while the restrained tower pattern appears to capture the frequency response with a difference amounting to only 4% along the *x* direction, but does not detect the second mode in the *y* direction.

Table 3. Comparison between the experimental and numerical results.

		Experimental	Numerical			
		f (Hz)	f (Hz)		MAC (%)	
Mode	Axis	OMA	FT	RT	FT	RT
1st	x	0.69	0.37	0.57	99	98
	y	0.76	0.37	0.72	99	99
2nd	x	2.28	1.65	2.20	95	94
	y	2.35	2.02	–	91	–

In Figures 8b and 8c, the numerically modelled (FT and RT) and the experimental modal shapes are compared in terms of flexural displacements, u_f, normalised with respect to the displacement at $z = 56$ m, corresponding to the maximum instrumented level on site. Visual analysis of the plots shows that the experimental modal shapes associated with the first modes are intermediate between those reproduced by the free tower (FT) and the restrained tower (RT) patterns. The good agreement between the experimental and both numerical modal shapes is confirmed by the values of the Modal Assurance Criterion (MAC) listed in Table 3.

It is worth remembering that the closer to 100% the MAC value, the better the approximation of the numerical results to the experimental data.

Because under strong-motion earthquakes the restrained model of the tower might become unrealistic, due to a likely progressive loss of restraint exerted by the lateral buildings, the dynamic nonlinear analyses were carried out considering both tower patterns, that is, free and restrained.

5 THE INPUT MOTION

The seismic safety of the Carmine Bell Tower was assessed under a reference input motion obtained from the E–W component of the accelerogram recorded during the Irpinia 23.XI.1980 earthquake (M_w 6.9) at the Tricarico seismic station (see Figure 1c), located on a tuff rock outcrop (Pacor et al., 2011). The signal was scaled to a peak ground acceleration PGA = 0.176g, estimated through the empirical correlation of Faenza and Michelini (2010) as corresponding to the local intensity $I_{MCS} = 8$ experienced in Napoli during the historical 1456 earthquake (see Figure 1c). A detailed study reported by de Silva (2016) shows that the spectral shape and intensity of the selected record are compatible with the ground motion and the damage induced by the same earthquake, as reported by Fracassi and Valensise (2007).

Considering that the selected input motion should represent the wavefield propagated from the source to the site in the 1456 earthquake, the recorded signal was assumed as propagating along a straight line joining the epicentre to the bell tower site (azimuth = 63°; see the map in Figure 1c). As a consequence, two proportional input motion components were obtained after projecting the signal, scaled to PGA = 0.176g, along N–S and E–W directions (Figure 9a).

The two projected acceleration time histories are plotted in Figure 9b. The E–W component (y-axis) is greater than that along the N–S direction (x-axis) because of the projection angle, 63° with respect to North. A band pass filter from 0.1 Hz to 25 Hz was applied to the natural records, in order to limit the frequency content to the dynamic range reliably reproduced by the soil–structure model. Following the conventional procedure adopted by FLAC3D, the input motions were simultaneously applied in terms of shear stress acting at the base of the model shown in Figure 6.

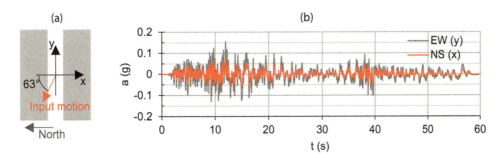

Figure 9. (a) Direction of the projection and (b) time histories of the acceleration of the input motion to the model.

6 NONLINEAR DYNAMIC RESPONSE OF THE SOIL–FOUNDATION–STRUCTURE SYSTEM

Figures 10 and 11 show the evolution of the plastic states, as well as of the tensile and compression strains, for the free and restrained tower patterns, respectively. The contours are associated with the time histories of the accelerations computed at the foundation level below the tower axis and 25 m away from it, that is, in free-field conditions.

In both cases, the structural yielding starts around $t = 10$ s, when the post-peak limit strain relative to compression for the tuff masonry ($\varepsilon_c = 0.4\%$; see Figure 7a) is first achieved. For the free tower (Figure 10), the consequent plastic deformations are concentrated at the base, and for the restrained structure (Figure 11) they are above the lateral contact with the adjacent buildings, that is, from $z = 19$ m to $z = 29$ m. Plastic points tend to diffuse with time, generating a sort of spread plastic hinge. The collapse of the tower is recognised in

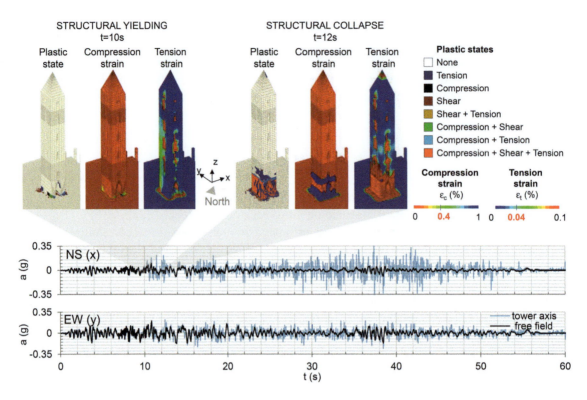

Figure 10. Plastic states and compression and tension strains corresponding to the yielding and collapse of the free tower model, associated to the time histories of the x and y accelerations.

both cases at around $t = 12$ s, when the strains in the entire cross section reach the compression limit, $\varepsilon_c = 1\%$, set for the tuff masonry.

Due to the different positions of the spread plastic hinges, the free structure and its foundation collapse together (Figure 10), while the foundation of the restrained tower attains the plastic state around $t = 36$ s (Figure 11), significantly later than the structural failure.

The N–S and E–W accelerations computed at the foundation level, and in correspondence with the tower axis, coincide with the free-field signals until the foundation masonry approaches the plastic state. In the elastic field, in fact, the kinematic interaction between soil and low-embedded foundations is expected to be negligible (Pais and Kausel, 1988). As plastic deformations occur, permanent displacements start to be accumulated in the foundation soil as well, with a consequent modification of the signal transmitted to the superstructure.

The Fourier spectrum amplitude of the acceleration time histories plotted in Figures 10 and 11 are shown in Figure 12 for the free and restrained towers. Compared to the free-field results, the signals recorded below the foundations generally show higher values, and the difference increases with the frequency. Despite the larger amplitude of the E–W (y) component of the input motion, the scatter is more significant along the N–S (x) direction, where the foundation is more deformable due to its geometry.

Further, in Figure 13a the translation of the foundation for both tower models (FT and RT) is compared to the free-field displacement of the ground at foundation level. The instant relevant to the occurrence of plastic deformations in the foundations ($t = 10$ s for the free tower and $t = 36$ s for the restrained model) corresponds to the triggering of permanent displacements, which occur higher in the N–S (x) direction. As expected, the presence of the lateral restraint inhibits the foundation movements, thus reducing the final displacement.

Figures 13b and 13c illustrate the time histories of the total and flexural displacements at the top of the tower. A form of harmonic oscillation with increasing amplitude can be recognised before $t = 10$ s, almost coinciding with the first peak of the accelerogram. After the displacement sheds the pseudo-harmonic motion, an uncontrolled growth testifies to the occurrence of failure. Coherently with the contours shown in Figures 10 and 11, the drift in the displacement starts for both models at around $t = 12$ s, that is, when plastic deformations affect the entire horizontal cross section of the tower.

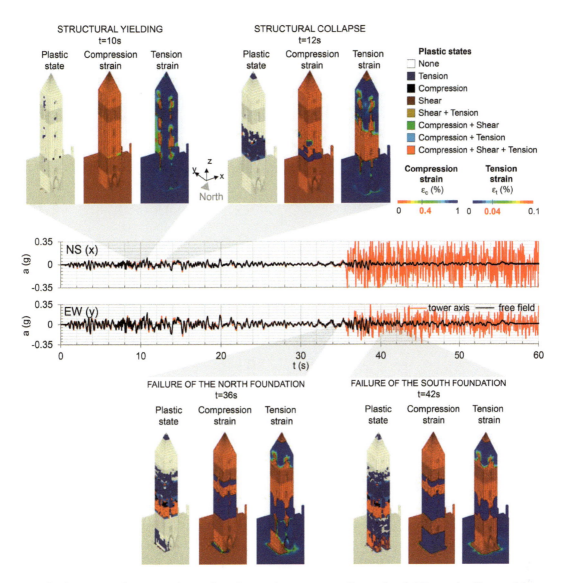

Figure 11. Plastic states and compression and tension strains corresponding to the yielding and collapse of the restrained tower model, associated to the time histories of the x and y accelerations.

In order to highlight the pre-failure behaviour of the tower, the deformed shapes of the axis at the instants corresponding to the maximum flexural displacement are shown along the x and y directions in Figure 14. According to the deformed shapes, the pre-failure oscillation corresponds to the first vibration mode. The flexural x displacements are slightly lower than those along y, consistent with the amplitudes of the input components. It is worthy of note that in a 3D model with a plastic constitutive model, if a yielding strain is reached in one direction, implying that a plastic state is achieved in the resistant section, the failure conditions are attained for both directions.

The translation of the foundation implies negligible rigid displacements in the structure, while the effects of the base rotation may be significant only for the free tower. In fact, as already observed, the movement of the foundation is reduced by the presence of the lateral restraint along the tower height.

7 SEISMIC BEHAVIOUR OF THE FOUNDATION

The behaviour of the soil at the foundation level was observed in detail in correspondence of points B and C, in the middle of the north and south footings, respectively. The attainment of plastic states for the soil is

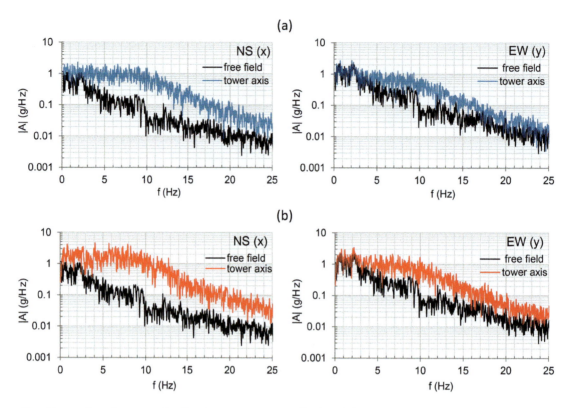

Figure 12. Comparison between the Fourier spectrum amplitude of the N–S and E–W accelerations computed at foundation level (in correspondence with the tower axis) and in free-field conditions for the two tower models: (a) free; (b) restrained.

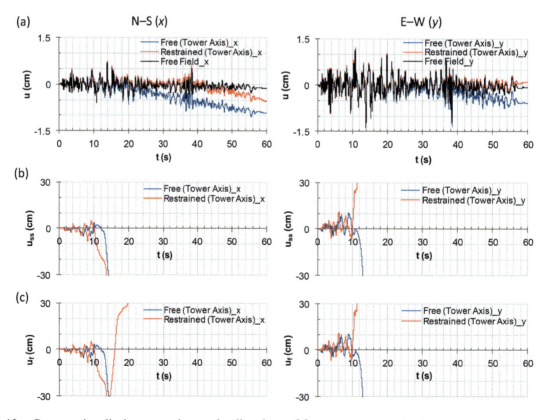

Figure 13. Comparative displacements in x and y directions of free tower vs restrained tower models: (a) foundation level; (b) total displacement at the tower top; (c) flexural displacement at the tower top.

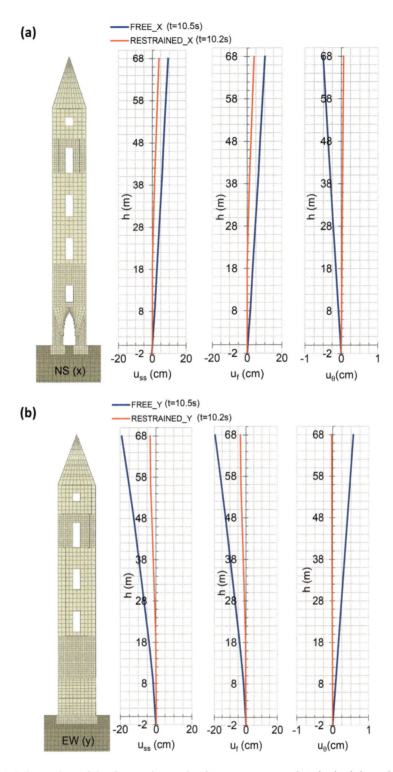

Figure 14. Structural deformation of the free and restrained towers computed at the incipient yielding times along the directions of: (a) N–S (x); (b) E–W (y).

shown in Figure 15 through the time histories of the principal effective stresses, σ'_1 and σ'_3, compared to the shear failure locus (inclined plane).

Both the control points below the free tower (Figure 15a) persist in the elastic field until the end of the earthquake, approaching the failure surface around $t = 10$ s, that is, when the tower fails. A sudden achievement of the plastic conditions is recognised for the restrained model (Figure 15b), corresponding to the foundation

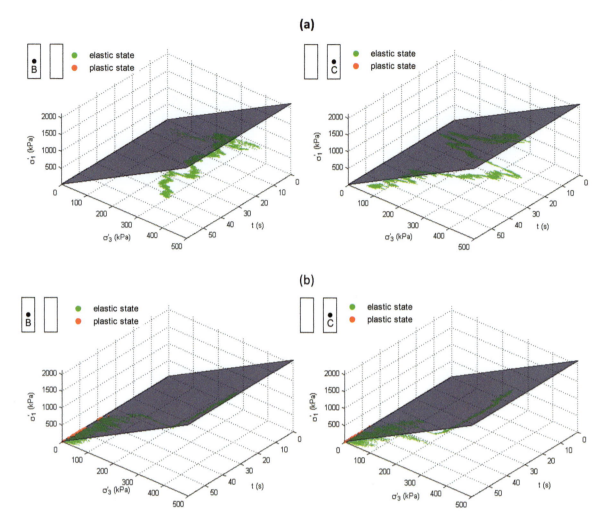

Figure 15. Time histories of the soil plastic state at points B (left) and C (right) below the foundations for: (a) the free tower; (b) the restrained model.

yielding (red points). The same results are plotted in Figure 16 through the time histories of the difference between the current maximum principal stress, σ'_1, and the failure value, σ'_{1f}, provided by the Mohr–Coulomb criterion for the current minimum principal compression stress, σ'_3, acting on the soil element. The pseudo-harmonic variation of the stress can be recognised until the free tower oscillates in the elastic field (grey lines), while the effect of the lateral constraints keeps the stress difference almost constant in the restrained model (black lines). A more irregular trend is shown as the foundation approaches the plastic state.

The time histories of the stresses computed in the foundation during the seismic analysis were integrated to obtain the axial force N (acting along the vertical axis z), the shear forces H_x and H_y (acting along the horizontal axes x and y), and the overturning moments M_x and M_y (acting, respectively, in the xz and yz planes), shown in Figures 17 and 18 for the free and restrained towers, respectively. As a result of the force equilibrium conditions, H_x and H_y, and M_x and M_y start from opposite values in x and from the same value in y. During the dynamic response, their values oscillate following loading-unloading cycles. The overturning moments attained by the free tower before the structural collapse ($t = 12$ s) are much higher than the corresponding values achieved by the restrained model. The presence of the rotational restraint, in fact, reduces the portion of the tower height involved in the overturning deformation and transfers the resisting section immediately above the restraint, at $z = 19$ m, that is, just where the spread plastic hinge occurs. As expected, after the spread of plastic strains in the basement of the free tower (Figure 17), the resisting forces and moments assume a steady value until the end of the earthquake. A sudden decompression of the foundation can be recognised around

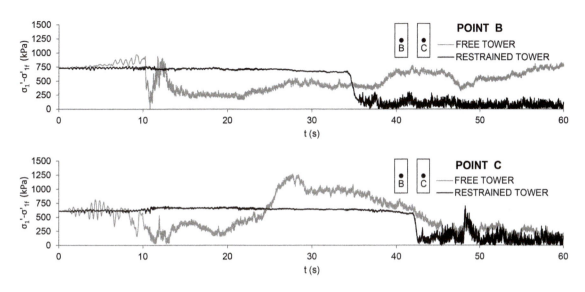

Figure 16. Time histories of the difference between the maximum principal stress and the failure strength at points B and C below the foundations.

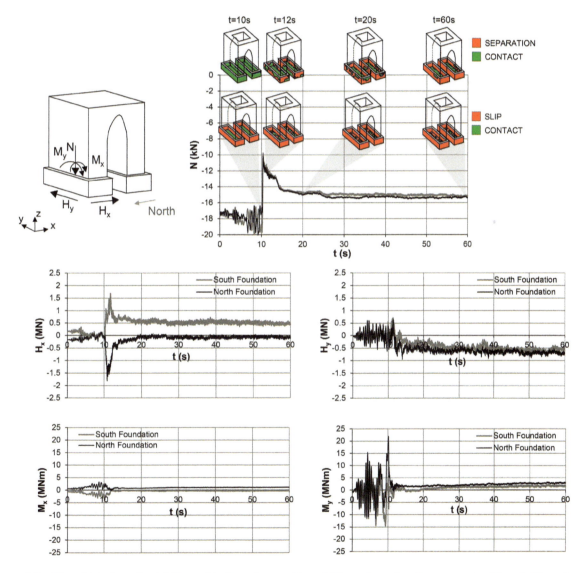

Figure 17. Time histories of axial force, N, shear forces, H_x and H_y, and overturning moments, M_x and M_y, acting on the north and south foundations of the free tower.

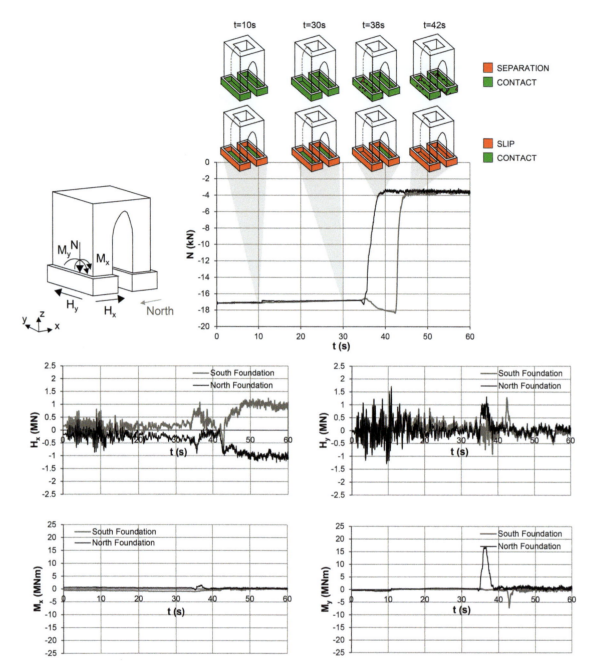

Figure 18. Time histories of axial force, N, shear forces, H_x and H_y, and overturning moments, M_x and M_y, acting on the north and south foundations of the restrained tower.

12 s, that is, when the interface attains plastic separation. The corresponding decrease of the axial force N is much more significant for the restrained model (Figure 18) than for the free tower (Figure 17).

Figures 17 and 18 also show the contours of the slip and separation points at the foundation–soil interface, for some particular instants of the motion with reference to the time history of N. As a result of the force equilibrium conditions, slippage occurs along the lateral interfaces due to the settlements. At the horizontal interface, slippage occurs when the value of the local shear load exceeds the shear strength, while separation is caused by a null value of the normal load. It is worth remembering that a Mohr–Coulomb model is associated with the interface, so that the shear strength decreases when the normal stress is reduced.

During the dynamic oscillation, slip and separation contours spread randomly for the free tower model (see Figure 17), following the irregular distribution of plastic deformations occurring at the base. A perfect contact is shown by the restrained tower model even after the structural collapse (Figure 18). As the decompression of

Table 4. Parameters of the foundation failure loci.

| | N_u | D | B | μ | ψ |
	kN	m	m	–	–
x	33626	2	3.5	1.06	0.52
y			10	0.79	0.40

the foundation starts, the consequent reduction of the shear strength according to the Mohr–Coulomb model induces slippage of the foundation base.

8 SEISMIC ASSESSMENT OF THE FOUNDATION THROUGH THE FAILURE LOCI

As a further investigation, the failure loci of the bell tower foundation were computed along both the x and y directions, as reported in Figures 19 (x direction) and 20 (y direction) with reference to the north footing.

Following the macro-element approach of Nova and Montrasio (1991), the equation for the foundation failure locus of a rigid rough foundation on an elastic/perfectly plastic cohesionless soil is defined by the expression:

$$\left(\frac{H}{\mu N_u}\right)^2 + \left(\frac{M}{\psi B N_u}\right)^2 - \left(\frac{N}{N_u}\right)^2 \left[1 - \left(\frac{N}{N_u}\right)\right]^{2\beta} = 0 \tag{3}$$

where:

- N is the normal load;
- H and M are, respectively, the resistant shear force and overturning moment, becoming $H = H_x$ and $M = M_x$ when the bearing capacity is computed along the x direction, or $H = H_y$ and $M = M_y$ along the y direction;
- N_u is the axial bearing capacity when H and M are null.

The value of β is usually set equal to unity, while μ and ψ are functions of the soil–foundation interface friction angle, δ, and of the foundation width, B, and embedment, D, as follows:

$$\mu = tg(\delta) + 0.72\frac{D}{B} \tag{4}$$

$$\psi = 0.35 + 0.30\frac{D}{B} \tag{5}$$

In this study, the parameter δ was set equal to the friction angle of the made ground ($\phi = 38°$), while D and B were inferred from the geometry of the foundation. The axial bearing capacity was computed from the well-known formula of Vesic (1975). The parameters used for the calculation are summarised in Table 4.

The histories of the vertical and horizontal forces, together with the overturning moments along x and y directions, are plotted against the associated failure loci in Figures 19a and 20a, respectively. The same histories are projected in Figures 19b and 19c and Figures 20b and 20c on two orthogonal cross sections of the domains in the H-N and M-N planes, respectively. Note that the shape of the domain varies between the two directions, due to the difference between the length and the width of the foundation. As a consequence, the shear strength is greater along x (compare Figures 19b and 20b), while a higher resistance to the overturning moments resulted along y (compare Figures 19c and 20c).

Figures 19 and 20 highlight that the load pattern for the restrained model approaches the failure locus, due to the reduction of N; rather surprisingly, safer conditions are predicted for the free tower. With respect to the failure conditions expressed in terms of stress and strain states resulting from the nonlinear dynamic analyses, the macro-element approach anticipates safer behaviour of the foundation. The difference can be

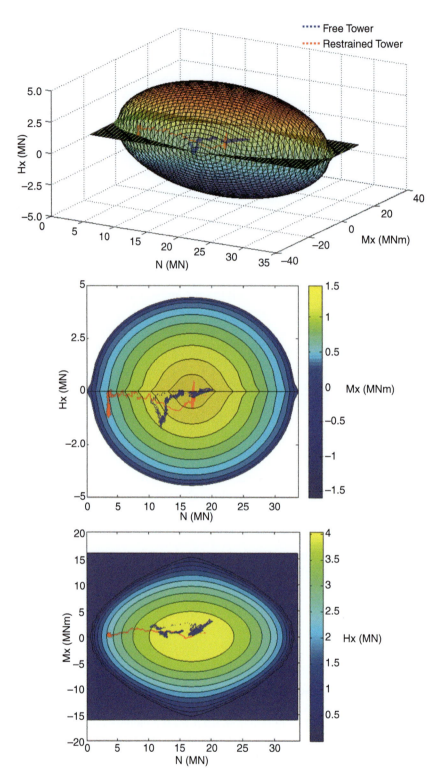

Figure 19. Time histories of N, H_x and M_x versus the failure locus of the north footing: (a) global view; (b) and (c) cross sections.

ascribed to the evolution of the resisting section, provided only by the finite difference model. Slippage and separation, in fact, reduce the contact area between soil and foundation, while in the macro-element approach the whole area of the footing is assumed to compute the resistance, independently of the actual history of the unloading-reloading cycles.

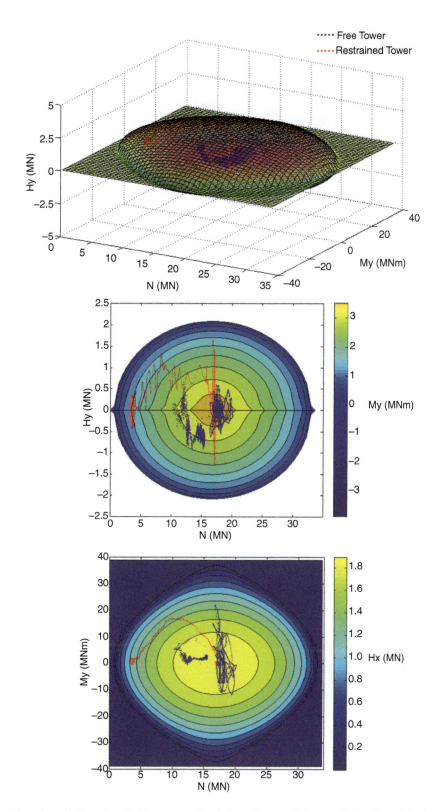

Figure 20. Time histories of N, H_y and M_y versus the failure locus of the north footing: (a) global view; (b) and (c) cross sections.

9 CONCLUSIONS

This chapter has illustrated the numerical procedure developed to model the soil–foundation–structure interaction of the Carmine Bell Tower in Napoli, an iconic case of a tall monumental building placed on a soft soil. An advanced 3D full dynamic model of the whole system was developed using FLAC3D code. Heterogeneity and nonlinearity of both soil and construction materials were accounted for, on the basis of accurate experimental investigations on the subsoil and the structure. Two different types of lateral constraints were considered for the tower, namely free and partially restrained along its height. The restrained tower model, reproducing the contact with the surrounding buildings of the Carmine complex, was able to simulate the modal shapes of the structure detected experimentally. Seismic nonlinear SFS analyses were then carried out by considering an input motion representing the destructive earthquake that occurred in 1456 and propagating as vertically incident shear waves in the subsoil.

The time history of the accelerations predicted below the foundation is identical to the free-field motion as long as the behaviour of the foundation and the surrounding soil remains in the elastic field, while significant differences between the two ground motions can be recognised as the foundation starts to accumulate permanent displacements. In addition, the discrepancy is more evident along the x direction, where the foundation is more deformable due to its geometry. While it is typically claimed in the literature that the kinematic interaction is generally negligible for a low-embedded foundation, this example shows that modifications in the foundation input motion can be caused by the plastic behaviour of the soil–foundation interface. Due to the significant tower displacements, high stress concentration occurs in the soil below the foundation, and plastic strains become more significant with respect to the free-field motion.

The lateral constraint along the tower height was found to strongly affect the seismic response of the structure, because it changes the position and the time of occurrence of the plastic hinges developing in the structure.

Finally, following the approach of Nova and Montrasio (1991), the failure loci of the foundation in x and y directions were computed in order to assess the foundation safety during the seismic shaking. In this case, the interaction between the tower foundation and the soil due to slippage and separation, and, hence, reduction in the contact area, regulate the safety condition of the foundation in a different way with respect to the macro-element model, where the whole foundation area is considered in the interaction.

The Carmine Bell Tower is intended as a prototype study for the seismic safety of historical towers, highlighting the need for an interdisciplinary action strategy in order to simultaneously achieve the conservation and fruition of the historical heritage. However, the vicissitudes of reality depart significantly from the research advancements, as highlighted by the current state of the bell tower, shown in Figure 21a, compared to that at the beginning of our study (Figure 21b). In 2015, due to the corrosion of the oldest chains, an anchor plate fell down from the bell tower. Fortunately, there were no consequences for any people, but, since then, the bell tower has been covered with scaffolding, waiting to be restored. Unfortunately, the studies reported in this chapter and previous ones too have been totally ignored by the authorities managing the restoration.

10 SUMMARY

Recent earthquakes in Italy have damaged outstanding historical centres, most of them dating back to the Middle Ages. Many churches and bell towers were seriously damaged too, with great pain for the population, since these structures represent the historical, social and religious identity of the communities. The assessment and improvement of the seismic safety of a monumental tower requires an accurate analysis of the dynamic response of the soil–foundation–structure (SFS) system, still overlooked by current practice.

This chapter focused on the iconic case of the 68 m-high Carmine Bell Tower, located in the coastal area of Napoli, and founded on soft alluvial and volcanic soils. SFS interaction analyses of the soil–tower system, subjected to a seismic input representative of the historical earthquake which destroyed the original tower, were carried out by developing an advanced dynamic 3D model. The nonlinear behaviour of both soil and structural materials (tuff and brick) was implemented through suitable hysteretic or linear elastic-plastic constitutive models, the parameters of which were accurately calibrated on the basis of field and laboratory investigations.

Different types of lateral constraint of the tower were considered in order to account for the likely interaction of the structure with the adjacent buildings of the monumental complex. The lateral constraint imposed along the tower height was found to strongly influence the natural frequencies and the modal shapes of the structure; the restraining conditions also affected the nonlinear seismic response of the system by changing the position and the time of occurrence of plastic hinge in the structure and the foundation.

Figure 21. Carmine Bell Tower: (a) as of 2016; (b) before these latest repair works.

Finally, the foundation safety was assessed in terms of stress and strain states at the soil–foundation interface, as well as by comparing the resulting forces against the failure locus of the footing. This latter approach, overlooking the slippage and separation between the tower foundation and the soil (hence, the reduction in contact area), was found to over-predict the safety conditions of the foundation.

ACKNOWLEDGEMENT

This work was carried out as part of WP5 'Soil–Structure Interaction' of the sub-project on 'Earthquake Geotechnical Engineering', in the framework of the research programme funded by Italian Civil Protection through the ReLUIS Consortium.

REFERENCES

Afan de Rivera, C. (1840). Tavola di riduzione dei pesi e delle unità di misura delle due Sicilie in quelli statuti dalla legge de' 6 Aprile 1840. http://www.storiamediterranea.it/portfolio/tavola-di-riduzione-dei-pesi-e-delle-misure-delle-due-sicilie/

Augenti, N. & Parisi, F. (2010). Constitutive models for tuff masonry under uniaxial compression. *Journal of Materials in Civil Engineering*, 22(11), 1102–1111.

Bommer, J., Douglas, J. & Strasser, F. (2003). Style-of-faulting in ground-motion prediction equations. *Bulletin of Earthquake Engineering*, 1(2), 171–203.

Celano, C. (1856). Notizie del bello, dell'antico e del curioso della città di Napoli. *Napoli, Stamperia Flaviana*.

Ceroni, F., Pecce, M., Voto, S. & Manfredi, G. (2009). Historical, architectural and structural assessment of the bell tower of Santa Maria del Carmine. *International Journal of Architectural Heritage*, 3(3), 169–194.

Ceroni, F., Pecce, M. & Manfredi, G. (2010). Seismic assessment of the bell tower of Santa Maria Del Carmine: Problems and solutions. *Journal of Earthquake Engineering*, 14(1), 30–56.

de Silva, F. (2016). Dynamic soil-foundation-structure interaction for masonry towers: The case study of Carmine bell tower in Napoli (PhD thesis). University of Napoli "Federico II".

de Silva, F., Ceroni, F., Sica, S., Silvestri, F. & Pecce, M.R. (2015). Effects of soil-foundation-structure interaction on the seismic behaviour of monumental towers: The case study of the Carmine Bell tower in Naples. *Rivista Italiana di Geotecnica*, Special Issue "Il ruolo della geotecnica nella salvaguardia dei monumenti e dei siti storici", 49(3), 7–27.

de Silva, F., Ceroni, F., Sica, S., Silvestri, F. & Pecce, M.R. (2017). Non-linear analysis of the Carmine bell tower under seismic actions accounting for the soil-foundation-structure interaction (in preparation).

Evangelista, L., de Silva, F., d'Onofrio, A., Di Fiore, V., Silvestri, F., Scotto di Santolo, A., Cavuoto, G., Punzo, M. & Tarallo, D. (2016). Application of ERT and GPR geophysical testing to the subsoil characterization of cultural heritage sites in Napoli (Italy). *Measurement* (in preparation). doi:10.1016/j.measurement.2016.07.042

Faenza, L. & Michelini, A. (2010). Regression analysis of MCS intensity and ground motion parameters in Italy and its application in ShakeMap. *Geophysical Journal International*, 180(3), 1138–1152.

Fioravante, V., Abate, G., Giretti, D., Aversa, S., Boldini, D., Crespellani, T., Dezi, F., Facciorusso, J., Ghinelli, A., Grasso, S., Lanzo, G., Madiai, C., Massimino, M. R., Maugeri, M., Tropeano, G., Santucci de Magistris, F., Sica, S., Silvestri, F. & Vannucchi, G. (2013). Earthquake geotechnical engineering aspects of the 2012 Emilia-Romagna Earthquake (Italy). *Proceedings 7th International Conference on Case Histories in Geotechnical Engineering, Chicago, 29 April–4 May 2013* (Paper no. EQ-5, pp. 1–34). ISBN #1-887009-17-5 (CD); ISBN #1-887009-18-3 (Abstract volume).

Fracassi, U. & Valensise, G. (2007). Unveiling the sources of the catastrophic 1456 multiple earthquake hints to an unexplored tectonic mechanism in southern Italy. *Bulletin of Seismological Society of America*, 97(3), 725–748.

Galante, A. (1873). Guida Sacra della Città di Napoli, *Fausto Fiorentino, Napoli, 1967, Edizione Fibrebo, Napoli.*

Itasca Consulting Group Inc. 2015 – FLAC3D (Fast Lagrangian Analysis of Continua) Version 5.01. Minneapolis, MN.

Monaco, P., Totani, G., Barla, G., Cavallaro, A., Costanzo, A., d'Onofrio, A., Evangelista, L., Foti, S., Grasso, S., Lanzo, G., Madiai, C., Maraschini, M., Marchetti, S., Maugeri, M., Pagliaroli, A., Pallara, O., Penna, A., Saccenti, A., Santucci de Magistris, F., Scasserra, G., Silvestri, F., Simonelli, L., Simoni, G., Tommasi, P., Vannucchi, G. & Verrucci, L. (2012). Geotechnical aspects of the L'Aquila earthquake. *Geotechnical, Geological and Earthquake Engineering*, 16, 1–66.

Moscarella, P. (1589, and continued by *alii* until 1825). *Chronistoria del Carmine Maggiore di Napoli.*

Nova, R. & Montrasio, L. (1991). Settlements of shallow foundation on sand. *Géotechnique*, 41(2), 243–256.

Pacor, F., Paolucci, R., Luzi, R., Sabetta, F., Spinelli, A., Gorini, A., Nicoletti, M., Marcucci, S., Filippi, L. & Dolce, M. (2011). Overview of the Italian strong motion database ITACA 1.0. *Bulletin of Earthquake Engineering*, 9(6), 1723–1739.

Pais, A. & Kausel, E. (1988). Approximate formulas for dynamic stiffnesses of rigid foundations. *Soil Dynamics and Earthquake Engineering*, 7(3), 213–227.

Rayleigh, Lord. (1945). *Theory of sound*, Vol. 1. New York, NY: Dover Publications.

Rovida, A., Camassi, R., Gasperini, P. & Stucchi, M. (Eds.) (2011). *CPTI11, the 2011 version of the parametric catalogue of Italian earthquakes*. Milan, Italy: Istituto Nazionale di Geofisica e Vulcanologia. http://emidius.mi.ingv.it/CPTI

Stewart, J.P., Lanzo, G., Aversa, S., Bozzoni, F., Dashti, S., Di Sarno, L., Durante, M.G., Simonelli, L.A., Penna, A., Foti, S., Chiabrando, F., Grasso, N., Di Pietra, P., Franke, K., Reimschiissel, R., Young, B., Galadini, F., Falcucci, E., Gori, S., Kayen, R., Kishida, T., Mylonakis, G., Katsiveli, E., Pagliaroli, A., Giallini, S., Pelekis, P., Psycharis, I., Vintzilaiou, E., Fragiadakis, E., Scasserra, G., Santucci de Magistris, F., Castiglia, M., Fierro, T., Mignelli, L., Sextos, A., Alexander, N., De Risi, R., Sica, S., Mucciacciaro, M., Silvestri, F., d'Onofrio, A., Chiaradonna, A., de Silva, F., Tommasi, P., Tropeano, G. & Zimmaro, P. (2016). *Engineering reconnaissance following the 2016 M6.0 Central Earthquake: Ver 2*. Geotechnical Extreme Events Reconnaissance (GEER) Association. doi:10.18118/G61S3Z

Vesic, A.S. (1975). Bearing capacity of shallow foundations. In: Winterkorn, H.F. & Fang, H.-Y. (Eds.), *Foundation Engineering Handbook*. New York, NY: Van Nostrand Reinhold.

The Leaning Tower of St. Moritz: A structure on a creeping landslide

A.M. Puzrin
Institute of Geotechnical Engineering, ETH Zürich

ABSTRACT: The 13th century Leaning Tower of St. Moritz, is located in the historic center of the famous Swiss ski resort town in the compression zone of a 10 million m^3 Brattas landslide. Over the hundreds of years, this slowly creeping landslide, which is blocked by a rock outcrop below the tower, had damaged the adjacent St. Mauritius church to such an extent that it had to be demolished already in 1893 due to dangerous differential settlements and cracks. The fact that the 33 m tall tower has survived its 5.40 degrees downslope inclination should not be taken for granted: this is an outcome of a century long effort by a number of outstanding Swiss engineers who came up over the years with original stabilization solutions. This chapter explores the history of the Leaning Tower of St. Moritz, describes its stabilization attempts and pays tribute to some extraordinary minds behind them.

1 INTRODUCTION

The historic Leaning Tower of St. Moritz (Figure 1) is located 50 m uphill from the Kulm Hotel in the center of the town. This luxury hotel was built in 1856 by the Swiss tourist industry pioneer Johannes Badrutt, mainly for English tourists, whom Badrutt attracted for the first winter holidays by betting that if they did not come

Figure 1. The Leaning tower of St. Moritz.

back home suntanned, he would pay for their trip. Needless to say, he won the bet and the winter tourism was born (e.g., Margadant and Maier, 1993). In 1879, the first electric bulb in Switzerland was lit in the Kulm Hotel dining hall. It was also the first hotel in Switzerland to boast a telephone, WC, hydraulic lifts and air heating, as well as winter sports such as curling and skeleton. Johannes Badrutt has been widely recognized and celebrated for these achievements, but there has been one more indication of his genius, which went totally unnoticed: namely, the choice of the hotel location, where not a single structural crack has been detected over the entire 160 years of its history. In contrast, if he built it just 50 m uphill from its present location, by today it would have to be demolished and rebuilt at least twice, if the owners still had the courage and finances to do so. All the historic structures above the hotel were subjected to slowly accumulating ground displacements, and eventually had to be demolished. Except one – the Leaning Tower of St. Moritz. This is its story.

2 THE BRATTAS LANDSLIDE

The Brattas-Fullun landslide, which constitutes the major factor for the special geotechnical conditions of the Leaning Tower, is located on the northern slope above the village of St. Moritz (Sterba *et al.*, 2000). It is composed of a 600 m wide clastic flow bounded on both sides by almost parallel shear surfaces (Figure 2, top).

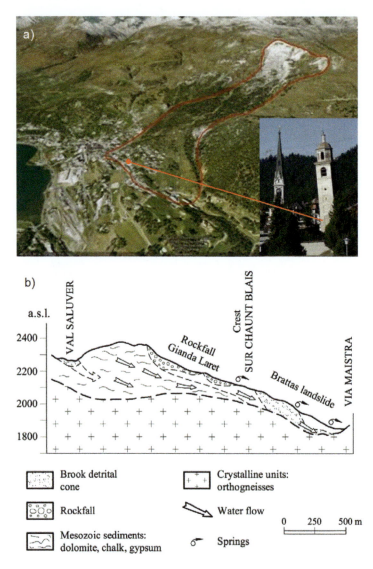

Figure 2. The St. Moritz-Brattas Landslide (top) aerial photo (after Google); (bottom) the geology (Müller and Messina, 1992).

The detachment zone is located on the southern edge of the terraced surfaces of the Val Saluver at an altitude of 2400 m a.s.l., and the area stretches over a horizontal distance of 1.5 km to a lower altitude of 600 m with the average inclination of about 20°. The clastic flow consists of two parts (Figure 2, bottom), with some geological evidence of a rock outcrop at the boundary between them. The upper zone, which extends from the detachment zone between Sass Runzöl and Sass da Muottas to the crest at an altitude of approximately 2100 m, is composed of a rockfall with boulders reaching 2–3 m in diameter. The lower 600–700 m long zone, which is the actual Brattas-Fullun landslide, is composed of a 17–23 m thick silty soil matrix with boulder inclusions, whose downhill movement is constrained at its foot by a rock outcrop. The landslide area is partially wooded at the top and has been developed for construction purposes in its lower portion. Until 2006 the movement has only been measured in the developed areas, reaching 5 cm per year above the construction zone and gradually tapering off to zero near the Via Maistra at the lower landslide edge (Figure 3).

The rockfall event is prehistoric in nature and has been the subject of many investigations by Swiss geologists (e.g., Cornelius, 1912; Heim, 1932; Schlüchter et al., 1982; Schlüchter, 1988; Müller and Messina, 1992; Schlüchter and Helfer, 1991). The large-scale geologic situation of the area resulted from Mesozoic sediments of the Bernina Nappe being pushed over the crystalline rock of the Err Nappe, subjecting all the layers in this region to high tectonic stresses and forming a lithologically very inhomogeneous and strongly weathered weak zone with liassic layers of clay-rich schists, which facilitate formation of slip surfaces.

The hydrological conditions further contributing to the slope instability: the saturation of the slope is due not only to frequent precipitation (average of 1000 mm per year) but also to the permanent seepage from the Schlattain Brook and periodic snow melt from the terraced surfaces of the Val Saluver (Figure 2b), verified through tracer tests by Müller and Messina (1992). Various deep aquifers were observed in the landslide, which create independent water tables in the slope formed by fossil soil layers overlain by later slides. Radiocarbon dating of wood and peat samples (Schlüchter 1988) enabled various movement events in the slope to be identified with the most recent one being approximately 700 years old.

The Leaning Tower is located at the foot of the Brattas slope north of the Kulm Hotel, above the Via Maistra (Figure 3) with its foundation based in the 15 m thick sliding mass. The underlying rock is highly deformed (Serizite-Muskovite gneiss and schists) and rises towards the Via Maistra to the ground surface, forming a rock ridge against which the creeping mass comes to a halt creating a compression zone with the creep movement of approximately 1 cm per year.

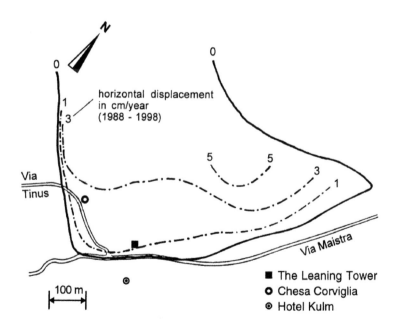

Figure 3. Average horizontal displacement rates 1988–1998 at the bottom of the landslide (after Tschudi and Angst, 1999).

3 THE LEANING TOWER AND THE STABILIZATION OF THE LANDSLIDE (ALBERT HEIM, 1899)

The bell tower of the St. Mauritius church was originally built in Romanesque style in the 13th century, approximately 100 years after the construction of the church (e.g., Haefeli, 1974; Margadant and Maier, 1993). In the course of the following centuries its height was increased, first in the 16th century, accompanied by installation of the chimes and of the clock. The top floor in the late Renaissance style was probably added in the 17th century. Existing records show that in 1797 the tower was displaced by an earthquake, raising concerns about its stability. In 1890, due to the continuously growing inclination, it was decided to decrease the overturning moment by removing the chimes, which were installed in a specially constructed wooden frame and exhibited in the vicinity of the church. Curiously, while the tower has been inclined almost exactly downslope, the adjacent St. Mauritius church, which was located above the tower, was inclined upslope. Maillart (1931) explained this phenomenon by a possible local heave of the compressed sliding mass between the church and the tower. In 1893, differential displacements of the church reached the level seriously endangering its structural stability and the church was demolished. In 1897 it was also considered to demolish the tower, but thanks to the very good quality of its masonry it was decided to leave the tower in its inclined state (estimated at that time as 2.5°–3.0°, Wullimann and Schneller, 1989). Concerned, however, about the continuing displacements of the area and inclination of the tower, the St. Moritz community invited in 1899 Prof. Dr. Albert Heim to look into their potential reasons and to suggest possible mitigation measures.

The choice of the expert was probably not difficult, since Professor Heim (1849–1937) was already at that time one of the top world geologists, considered to be one of the founding fathers of the Engineering Geology. As recently as 2012, the Joint Technical Committee 1 "Natural Slopes and Landslides" which is a scientific committee created by three of the main societies in the field of Geotechnics: ISSMGE (International Society of Soil Mechanics and Geotechnical Engineering), ISRM (International Society of Rock Mechanics) & IAEG (International Association of Engineering Geology) established an International Lecture dedicated to Albert Heim, that would take place every four years during the International Symposium on Landslides.

Albert Heim became full professor of technical and general geology at the ETH Zurich (at that time Polytechnic Institute) in 1872, at the age of twenty-three, which was remarkable even for that time (e.g., Brockmann-Jerosch et al., 1952). Three years later he also became a professor at the University of Zurich and until his retirement in 1911, he gave hundreds of lectures, including "General Geology", "The Geology of Switzerland" and "The Prehistory of Man". His field trips in Switzerland and abroad were legendary among his students and colleagues.

The main focus of Albert Heim's monographic and cartographical works was on the origin of the Alps and on their geology. His innovative use of scientific drawing and photography (Figure 4) for geological documentation has profoundly influenced future generations of geologists. Prof. Heim authored numerous geological reports, e.g., for railway construction projects like the Simplon Tunnel or accidents such as the Vorstadt disaster in Zug (1887). He was involved in various commissions and societies, e.g., as the longstanding President of the Swiss Geological Survey. Beyond his geological interests, he was also an expert in cynology, championing among other breeds the famous Bernese Mountain Dog.

In his expert report of 1899, Prof. Heim established that the main reason behind the continuously increasing displacements of the Leaning Tower of St. Moritz and surrounding structures was a permanent landslide and proposed to reduce its velocity by installing drainage system in the upper part of the landslide. Implementation of these mitigation measures was recognized by Maillart (1931) und Haefeli (1974) as critical for preventing the large-scale deformations and improving the general stability of the slope. It could not, however, stop the displacements completely, as is seen from the results of the continuous tower inclination measurements undertaken starting from 1908 (Figure 5). This is hardly surprising, since the ground water comes into the sliding mass from a very broad area (see tracer tests by Müller und Messina 1992) and cannot be fully drained using a local drainage system. By 1928, inclination of the tower increased to 4.4° (7.7%) and the danger of the eventual tower collapse became imminent (Maillart, 1931).

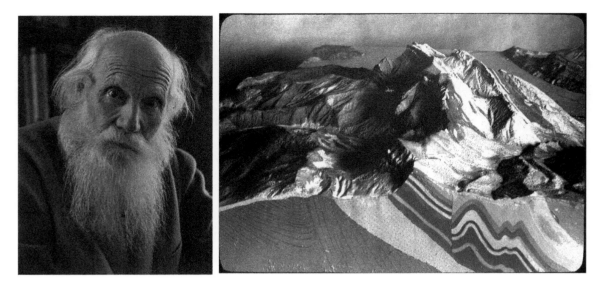

Figure 4. Albert Heim and an example of one of his geological profiles (source: ETH – Library).

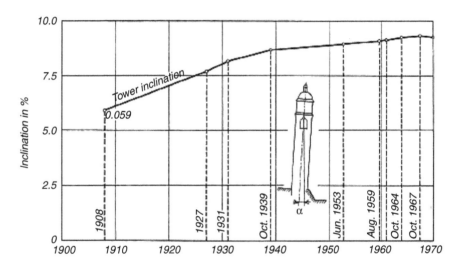

Figure 5. Inclination of the tower in % between 1908 and 1967 (after Haefeli, 1974).

4 THE FIRST STABILIZATION OF THE TOWER (ROBERT MAILLART, 1928)

In 1927, the concerned municipality invited Engineer Robert Maillart to explore the possibility of the tower stabilization. Robert Maillart (1872–1940) was a famous Swiss civil engineer and bridge designer. His groundbreaking use of structural reinforced concrete radically changed both the architecture and engineering of bridge construction (e.g., Marti, 1996). In 1991, his Salginatobel Bridge built in 1930 (Figure 6) was declared an International Historic Civil Engineering Landmark by the American Society of Civil Engineers. In 2001, the British journal *Bridge – Design and Engineering* voted Salginatobel Bridge as "the most beautiful bridge of the century". Maillart's designs of the beamless floor slabs and mushroom ceilings for industrial buildings (Figure 6) are another example of his artistic and engineering ingenuity.

Robert Maillart completed his Civil Engineering studies in 1894 at the ETH Zurich, and after working for a few years in industry, established in 1902 his own company *Maillart & Cie*. The company was engaged in large-scale projects in Switzerland and Europe and eventually brought him and his family in 1912 to Russia.

Figure 6. Robert Maillart, his Salginatobel bridge and the mushroom slab (source: Marti, 1996).

Tragically, he was trapped there during the WWI and the revolution, his wife died and his company was nationalized. Escaping in 1918 back to Switzerland as a bankrupt widower with three children, he mastered, however, a remarkable comeback leading to his best designs and projects.

For the Leaning Tower, Maillart established that with the lower edge of the foundation experiencing the load of 700 kPa, this did not only contribute to the tower inclination, but could also bring it close to the bearing capacity failure (Maillart, 1931). It was decided to implement stabilization in two steps (Figure 7): (i) decreasing the acting earth pressures upslope, and (ii) increasing the foundation area downslope. The work was carried out by the firm Prader & Cie (R. Coray). The first step was accomplished by excavating a trench behind the tower (Figure 7), which was later supported by concrete plates. In the second critical step, a concrete base was constructed starting from 3 m above the ground level, connected to the existing masonry and transferred the load to the extended foundation (Figure 7). This allowed do shift the eccentricity of the resultant force from 1.23 m downslope to 0.26 m upslope, reducing the load on the lower edge of the foundation from 700 kPa to 200 kPa (without the wind load) and eliminating the imminent bearing capacity failure.

Needless to say, the task was challenging since it involved excavation into soil and masonry at the overstressed downslope part of the tower. In order to avoid potential collapse during stabilization works, the tower was pulled back by 16 tension cables connected to the top openings in the tower and anchored back in the heavy concrete blocks. Additionally, the construction of the downslope concrete base, excavation and extension of the foundation was carried out in five narrow stages. During the works and after their completion in November 1928 the inclination rate increased (Figure 5), which is hardly surprising, since the new extended foundation required certain displacement to mobilized the contact stress. After 1939, however, significant slowing down was observed, which bought the tower another 30 years of life.

5 THE SECOND STABILIZATION OF THE TOWER (ROBERT HAEFELI, 1968)

From 1961 inclination of the tower started accelerating (Figure 5) and the municipality of St. Moritz felt it necessary to invite an expert opinion – this time of Professor Robert Haefeli. Robert Haefeli (1898–1978) was not only the pioneer of the geotechnical research in Switzerland but also one of the founders of the international Soil Mechanics (the Third World Congress of the still young International Society of Soil Mechanics and Foundation Engineering took place in 1953 in Zurich). Professor Haefeli was also one of the founders of the flagship international journal *Geotechnique*. His contribution was widely acknowledged during the celebration of the 60th Geotechnique anniversary in 2008 (Burland, 2008). Remarkably, his achievements in Snow Mechanics and Glaciology are considered in these disciplines to be as significant as in Soil Mechanics.

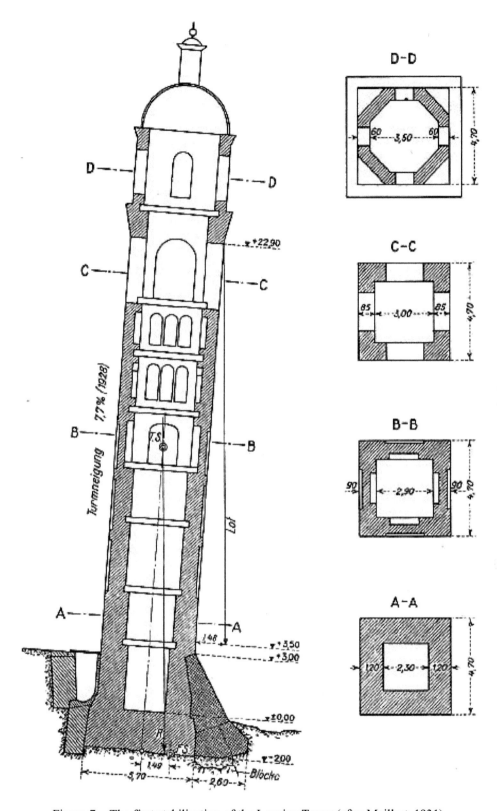

Figure 7. The first stabilization of the Leaning Tower (after Maillart, 1931).

It was largely due to Haefeli that a major expedition was organized to study the Greenland ice sheet. He was one of the co-founders of the Swiss Institute for Snow and Avalanche Research, and in 1954–1957 he was the president of the International Commission for Snow and Ice. His passion for rocks and soils, snow and ice went far beyond his research: Robert Haefeli loved the Alps and the open-air life, rock climbing and skiing.

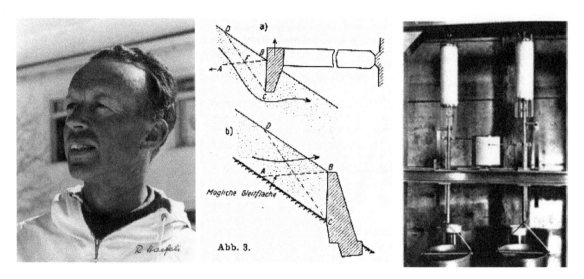

Figure 8. Robert Haefeli (after Burland, 2008), his theory of landslide pressure (center, after Haefeli, 1944) and snow experiments in 1936 (right, source E. Bucher, ETHZ).

After graduating in Civil Engineering from ETH Zurich, Robert Haefeli worked for a few years in Spain, where he was engaged in waterpower development schemes. His interests in soil, snow and ice mechanics were inspired by his work on the construction of dams in the Alps. In 1935 he was invited to establish the first soil mechanics section (including a laboratory) at ETH Zurich, where he carried out fundamental studies on the shearing resistance of soil, snow and ice. His doctoral thesis, published in 1939, was entitled "Snow mechanics with reference to soil mechanics." He was appointed a professor at ETH Zurich in 1947, and lectured on soil and snow mechanics together with avalanche mitigation and preventive measures (Figure 8). He was extremely active in research and his formulas for calculations of the landslide pressures are still extensively used in geotechnical design (Haefeli, 1944).

The immense value of his achievements becomes only more evident against the background of tragically lengthy periods of depressive ill health, which Haefeli suffered over the years and which forced him to leave ETH Zurich in 1953. He continued, however, to be active as a private consultant and it was in this capacity that the St. Moritz municipality invited in 1967 his expertise on the next stabilization of the Leaning Tower, which at that time increased its inclination to 5.37° (9.4%).

In his assessment of the Leaning Tower conditions, Prof. Haefeli calculated stresses in the structure and in the soil, caused by both the tower inclination and the wind load, which had not been considered before. His conclusion was that the stresses in the tower masonry were far from reaching a failure state. The stresses in soil have increased since the last stabilization of Mallart in 1928 from 200 kPa to 290 kPa, and have also not yet reached the bearing capacity. The main concern was, therefore, the continuing inclination of the tower with increasing average rate (from 0.02% per year in 1939–1961 to 0.05% per year in 1961–1964), which could, eventually, bring the stresses to the critical values. Thus, the solution was needed to slow down the inclination rates and, if possible, to find a way to control the contact stresses.

Professor Haefeli proposed to carry out the stabilization in three stages (Haefeli, 1974). In Stage I, the foundation of the tower was anchored with the help of four 300 kN, 20 m long Duplex anchors in the soil behind the tower (Figure 9). The main goal of this stage was to keep the tower stable during the works for the following stages. Stressing the anchors has caused 0.01° rotation in the upslope direction (opposite to the tower inclination), confirming their effectiveness. Stage II included installation of the drainage system below the tower foundations. Recognizing that the drainage installed in 1899 by Prof. Heim 150 m above the tower improved the slope stability, it was clear that it hardly affected the local conditions around the tower. To address that, three horizontal 25 m long, 131 mm diameter wells spread in a fan were drilled below the tower (Figure 9). The three drainage pipes had special filter perforation in the rear 15 m, bringing water into a collector. In spite of the fact that only one pipe brought a few drops ground water, the measure has proven to be rather effective causing the tower to rotate almost immediately backwards by 0.03°. A possible explanation

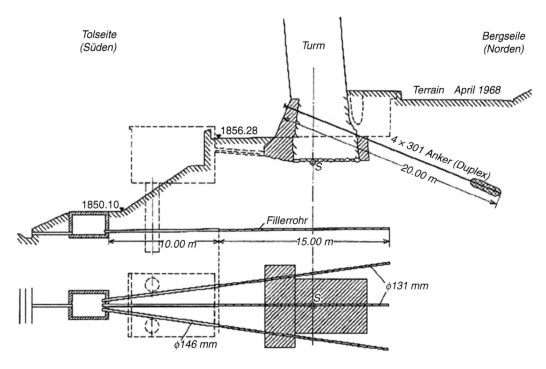

Figure 9. The second stabilization of the leaning tower (after Haefeli, 1974).

is that lowering one of the local ground water horizons caused increase of effective stresses, leading to the settlements behind the tower.

Although, the effectiveness of the first two stages caused the Stage III to be abandoned, it is still worth mentioning as an illustration of Prof. Haefeli's creative thinking. In order to be able to apply resisting moment to the tower foundation, he proposed to build a lever system (dashed lines in Figure 9). The principal function of this system is illustrated in Figure 10. Prof. Haefeli suggested that applying the moment opposite to the overturning moment could not only decrease the stresses in soil and structure but also affect kinematics of the inclination by triggering the soil creep in the opposite direction. Furthermore, in his paper (Haefeli, 1974) he presented a comparison between the Leaning Tower of St. Moritz and the Leaning tower of Pisa and suggested that the lever system can be also used for stabilization of the Leaning Tower of Pisa!

6 THE THIRD STABILIZATION OF THE TOWER (H.-J. LANG, R. WULLIMANN, F. SCHNELLER, 1983)

Professor Haefeli continued to monitor the tower after the second stabilization. In 1973 the anchors, which apparently lost some force, were pre-stressed back to 300 kN. In 1976 due to his health problems Professor Haefeli had to terminate his consulting activity and the municipality contracted the Institute of Geotechnical Engineering, ETH Zurich (at that time Institut für Grundbau und Bodenmechanik, IGB) to continue the monitoring of the tower. The institute had grown from the soil mechanics section founded by Haefeli in 1935 and became independent in 1970 under the leadership of Professor Hans-Jürgen Lang.

Hans-Jürgen Lang (b. 1931) graduated in Civil Engineering at ETH Zürich and after a distinguished career in industry was appointed in 1968 the first Professor of Soil Mechanics and Geotechnical Engineering at the ETH Zurich. He headed the institute for 28 years through a very exciting and productive period, when apart from its extensive teaching and research activities the institute supported many major geotechnical projects in the country. Professor Lang has not only been an outstanding engineer but also a legendary teacher, admired by generations of ETH students for his interesting and well structured lectures. His classical textbook "Bodenmechanik und Grundbau", which has already survived 10 editions, is being used in many German-speaking universities and can be found on the shelves of all geotechnical consultants in Switzerland.

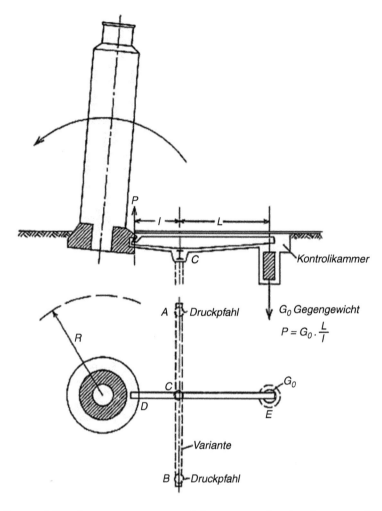

Figure 10. Proposal for a lever system to control the moments in a leaning tower (Haefeli, 1974).

Written independent of national codes, it focuses on basic mechanical principles of geotechnical problems and inspires engineers towards creative thinking.

The beginning of the IGB monitoring activity coincided with a rather alarming event: on 6th of May 1976 a Friaul earthquake of $M_S = 6.5$ took place in Northern Italy. The tower reacted with additional tilt of about 0.01°, i.e., almost a yearly value. Actually, during the second stabilization in April 1968 the tower reacted similarly to a small earthquake (Haefeli, 1974), but at that time the ground anchors had not yet been pre-stressed. Sensitivity of the tower to such relatively weak seismic loads became a concern and it was decided to review the stability of the tower based on the results of existing site investigations, inclinometer and piezometer measurements. Based on the limit analysis and finite element calculations it was concluded that the safety factor against bearing capacity failure below and outside the tower was as small as $FS = 1.04$ and could easily drop to unity as a result of relatively moderate increase in the ground water level (Wullimann and Schneller, 1989). It has, however, to be noted that the strength values adopted for these calculations were rather conservative, which probably explains why the tower could survive almost 2.5 times higher contact stresses before its first stabilization in 1928.

In any case, regardless of what was the true safety reserve, it was clear that with the gradually increasing tower inclination it would sooner or later become insufficient. And because the Stage III of the second stabilization proposed by Prof. Haefeli (Figure 10) had not been implemented, there was still no mechanism in place to control the stresses and inclination. The idea of the third stabilization proposed in 1982 by IGB (Eng. Rudolf Wullimann and Prof. H.-J. Lang) and the engineering firm Edy Toscano AG (Eng. Fredi Schneller), however, differed radically from the Haefeli lever system and included the following components (Figure 11). First, the tower was underpinned by two reinforced concrete foundation walls carefully sank to the depth of

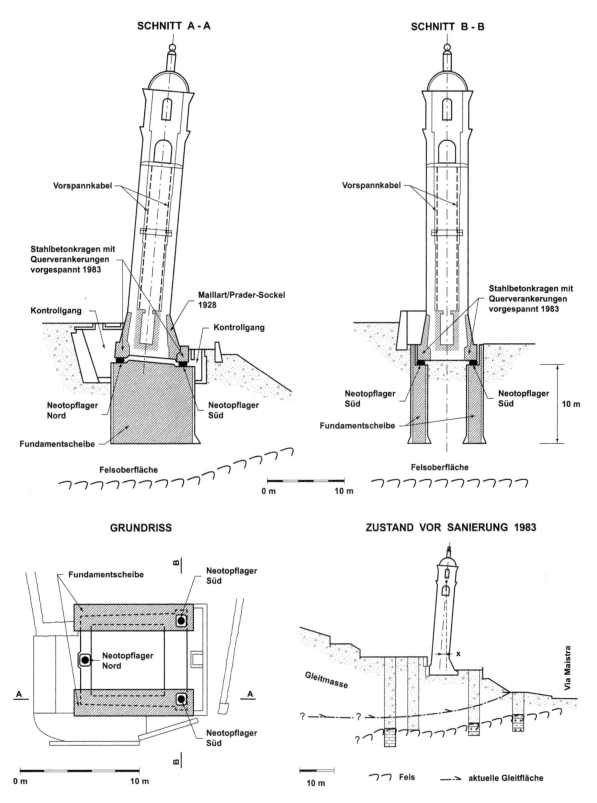

Figure 11. Schematic drawing of the third stabilization of the Leaning Tower of St. Moritz (information poster of St. Moritz community).

10 m, i.e., about 3 m above the sliding surface (Wullimann and Schneller, 1989). Next, pre-stressed reinforced concrete collars were placed in the area at the foot of the tower (Figure 12) to pre-stress the masonry of the foundation and bind together the original foundation and the footing extension constructed by Maillart in 1928. The tower weight (1264 t) was then lifted from the ground using hydraulic jacks and transferred onto the

Figure 12. Construction of the pre-stressed reinforced concrete collars at the tower footing.

Figure 13. Teflon bearing pad (left) and installation of the new plate during the 2005 tilt correction (right).

foundation walls through three Teflon bearing pads located between the collar beams and the foundation walls (Figure 13). Finally, the masonry of the tower was reinforced using vertical internal pre-stressing by means of the tension cables. Similar to the first stabilization in 1928, the safety of the tower during the construction works was enabled by the cables and ground anchors (Figure 14), with cables placed in the pipes to reduce temperature strains.

After the completion of construction works in 1983, the tilt of the tower was decreased by about 0.27°, bringing it 30 years back in time (nobody needs a "straight" leaning tower!). The engineering solution for the third stabilization left the Teflon bearing pads accessible allowing for the future corrections of the tower tilt by lifting its downslope edge and introducing additional steel plates into the bearing pads (Figure 13). Since 1983 the inclination of the tower has been carefully monitored by the St. Moritz municipality (Mr. Pietro Baracchi) using two independent measurement techniques: theodolite outside and Zenitlot inside of the tower (Figure 15). The measurements indicated that in 2005 and 2013 the tower approached its 1983 inclination (Figure 16),

Figure 14. Safety cables during the construction works for the first stabilization in 1928 (left, after Maillart, 1931) and the third stabilization in 1983 (right, after Wullimann and Schneller, 1989).

Figure 15. External theodolite (left) and internal Zenitlot (right) measurements.

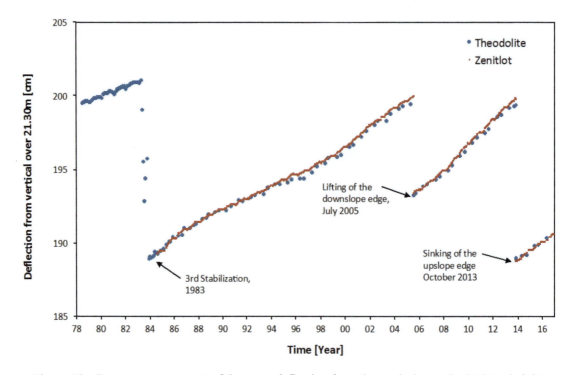

Figure 16. Recent measurements of the tower deflection from the vertical over the 21.30 m height.

which in 2005 was corrected by the engineering firm Edy Toscano AG (Eng. Dino Menghini) with the ETH Zurich support (Eng. Ivo Sterba and Dr. Dominik Hauswirth) by lifting its downslope edge and adding a plate into each of downslope bearing pads, and again in 2013 by lifting its upslope edge and removing a plate from the upslope pad.

7 LANDSLIDE ACCELERATION

Recent investigations of the landslide (Puzrin and Schmid, 2011; Schwager and Puzrin, 2014; Oberender and Puzrin, 2016) have provided some important insights into its mechanism. It has been established that after a long period of decreasing displacement rates, from the beginning of the 1990s the landslide started to accelerate (Figure 17). The "usual suspect" for the creeping landslide acceleration is an increase in precipitation. Indeed, the landslide displacements exhibit sensitivity to precipitation rate changes (e.g., in 2001–2006, when deviation of the precipitation rates from the average value is observed in Figure 17). Around the onset of acceleration, however, no precipitation increase could be observed, which was also confirmed by fairly constant groundwater table (Figure 18), measured using a piezometer pipe installed in a borehole about 37 m above the rock outcrop at the depth of 14 m. No rising in the groundwater table has been observed around the onset of acceleration in 1991. On the contrary, the groundwater table around 1991 seems to drop below the average level (with the average depth of 9.92 m), which appears to be the same both in the period of 1984–1996 (around the onset of acceleration) and during the entire measurement period of 1978–2010.

A more likely reason for the landslide acceleration could be the passive failure of soil at the bottom of the landslide, in the zone characterized by high compression strain rates (Figure 19) within the constructed area in the lower 200 m of the landslide in 1988–1998, close to the latitudinal center of the landslide. Puzrin and Schmid (2011) provide both theoretical and experimental evidence supporting the passive failure hypothesis.

Measuring in 2006–2010 for the first time displacement rates along the entire landslide length (close to the latitudinal center of the landslide) has shown (Figure 20) that the rates continue increasing upslope and can reach up to 0.5 m/yr (600–700 m above the Via Maistra and just below the foot of the rockfall). This is an order of magnitude higher than the maximum average displacement rates in the constructed zone and suggests that the compression zone along the Via Maistra (including the tower location) can be a subject of significant earth pressures.

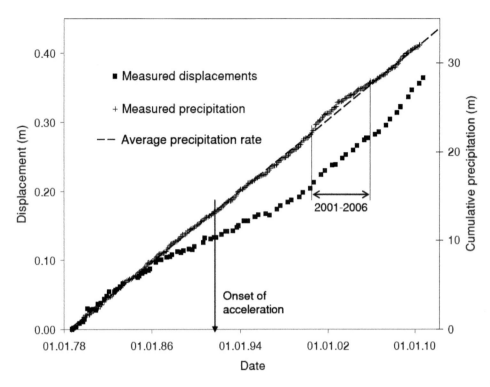

Figure 17. Displacements and cumulative precipitation in 1978–2010. Displacements were measured in the vicinity of the Leaning Tower (21.1 m away from the rock outcrop), in the direction of the slope gradient (after Puzrin and Schmid, 2011).

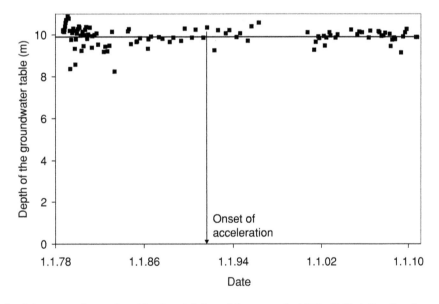

Figure 18. Depth of the groundwater level in the vicinity of the tower in 1978–2010 (after Puzrin and Schmid, 2011).

The changes in earth pressures were back-calculated from the changes in the shape of inclinometer pipes measured in 2008–2012 using the novel inclinodeformeter (Figure 21, after Schwager and Puzrin 2014). The average yearly increase in earth pressure in the compression zone was 0.3 kPa, an order of magnitude lower than the pressure increase outside this zone. At the same time, the strain rates in the compression zone are an order of magnitude higher than outside (Figure 19). This implies plastic flow at practically constant stresses, directly confirming the failure hypothesis.

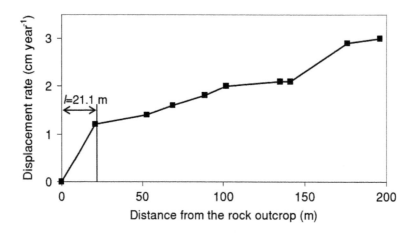

Figure 19. Average displacement rates within the constructed area in the lower 200 m of the landslide in 1988–1998 (after Puzrin and Schmid, 2011).

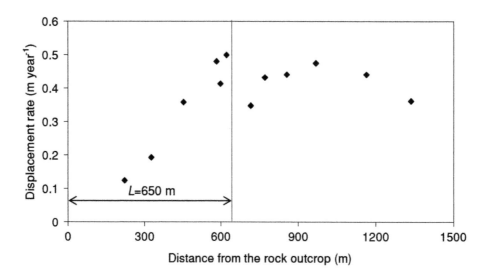

Figure 20. Average displacement rates in 2006–2010 above the constructed area. Displacements were measured using GPS device with the accuracy of 0.02 m (after Puzrin and Schmid, 2011).

8 MONITORING

The question, what exactly causes today the continuously increasing inclination of the tower, can be answered using the results of displacement measurements. The displacements and rotations of the tower have been occasionally measured since 1908 (Figure 22). From the beginning of the 1980s, the tower has been monitored on a regular basis using several complementing types of manual measurements. Consistent with other observations, these measurements show that in the last three decades the ground surface in the vicinity of the tower experienced acceleration of average downslope displacement rates (Figure 23): 6.5 mm/year in 1986–1996; 12 mm/year in 1996–2006 and 19.5 mm/year in 2006–2016. However, as is seen from inclinometer readings (Figure 24), displacement rates decrease with depth, resulting at the slip surface in about 10 mm/year (in 2006–2010). The tower foundation walls are 10 m deep and are likely to be effected by this non-uniform velocity field. Their rotation expresses itself as a differential settlement/heave at the four corners of the tower at the ground surface and leads to the increasing tower inclination. Using the measurement base distance of 21.3 m (Figure 23), the average tower deflection rate of 6 mm/year corresponds to a tilt rate of around 0.016°/year or about 3 mm per 10 m of the foundation wall depth, consistent with the inclinometer readings in Figure 24.

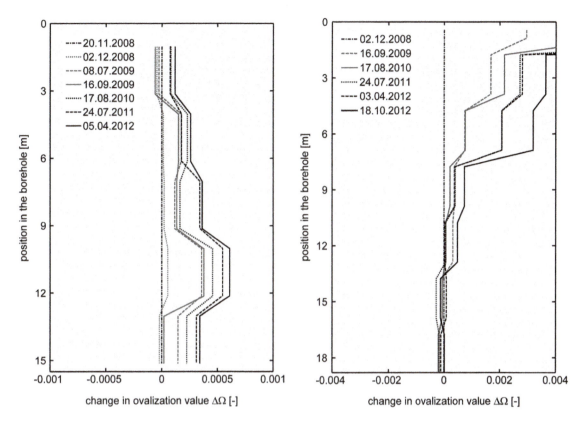

Figure 21. Evolution of the inclinometer pipe shape within the failed compression zone (left) and above the failed zone (right). Ovalization value Ω is calculated as the difference of the perpendicular pipe diameters normalized by the nominal outer radius of the inclinometer pipe R (after Schwager and Puzrin, 2014).

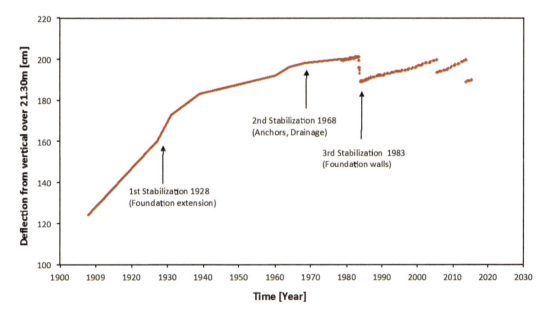

Figure 22. Approximate historic measurements of the tower deflection from the vertical over the 21.30 m height.

Acceleration of the landslide displacement rates caused by the soil failure in the compression zone has been closely followed by the tower inclination (Figure 23). In order to monitor these developments closer and to be able to react to extreme events, in addition to the manual measurements an automatic monitoring system was designed by the Institute of Geotechnical Engineering, ETH Zurich, and installed by Solexperts

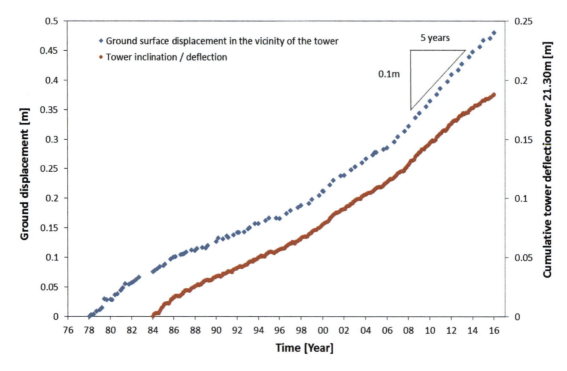

Figure 23. Ground displacements and cumulative inclination of the tower 1978–2016.

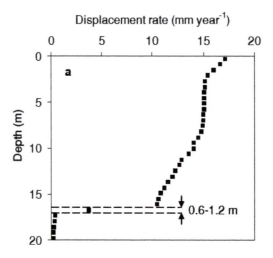

Figure 24. Average displacement rates in depth measured 2006–2010 by an inclinometer in the vicinity of the Leaning Tower (after Puzrin and Schmid, 2011).

AG (Figure 25). The system includes an inclination sensor (A) placed in a heated steel box to enable winter measurements; temperature and air pressure sensor (B); two in-place inclinometers (C1) and (C2) and a piezometer (D), all placed at the sliding surface level in boreholes outside the tower; and three accelerometers (E1 at the rock outcrop in the basement of the Kulm Hotel, E2 at the tower foundation and E3 at an upper floor level of the tower). All the data is accessible online and automatic text messages are sent when the measured values reach alarm values.

The system is extremely sensitive and could detect very small changes in tower inclination and pore water pressures caused by construction works in the vicinity of the tower. During the tilt correction in 2013, the inclination sensor (A) readings were successfully validated against the Zenitlot measurements (Figure 26). Vibrations of the tower, which were caused by a slight impact when the tower was lifted and brought back

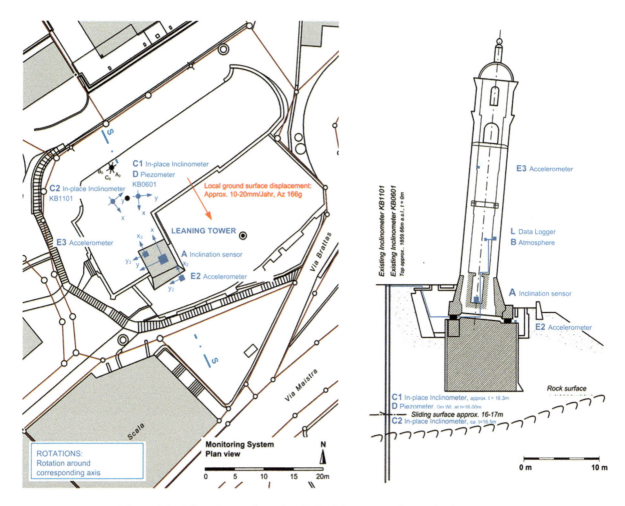

Figure 25. Plan view and section S–S of the automatic monitoring system.

into the contact with the bearing pad after a plate was removed, were detected by the accelerometer (E3) (Figure 27), allowing for assessment of the natural frequency of the tower. The 2016 earthquakes in Italy have also been detected by all three accelerometers and have been used for validating the site response spectra.

9 OUTLOOK

More than a hundred years of efforts by brilliant Swiss engineers have kept the Leaning Tower of St. Moritz alive, long after all other structures in its vicinity were demolished, rebuilt and demolished again. How to keep a historic tall narrow structure stable in a compression zone of a creeping landslide is not something a geotechnical engineer is being taught at school. The enormous, 10 million cubic meter landslide is not going to stop, on the contrary it has been currently accelerating, while the existing Teflon bearing pads have exhausted their capacity for changing their height by adding/removing plates. Furthermore, it has been established (Edy Toscano AG) that the existing solution does not satisfy the most recent norms for seismic stability.

Stochastic models of landslide displacements (Oberender and Puzrin, 2016) predict that the tower will reach its maximum historical inclination of 5.40 degrees again in about 10 years. Before this happens a new solution has to be found and implemented, including seismic isolation. Hopefully, the Leaning Tower of St. Moritz will continue challenging new generations of engineers for many years, decades and centuries to come!

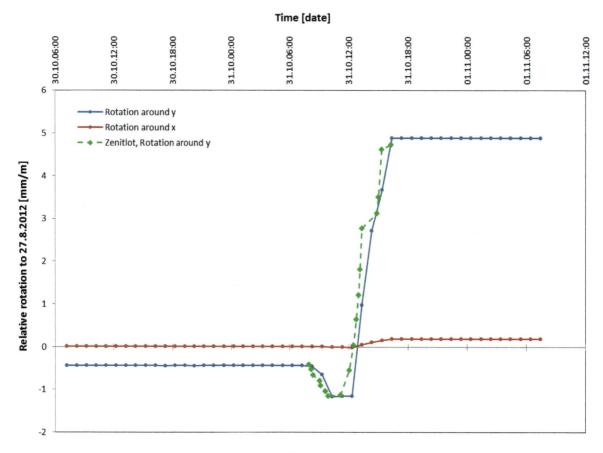

Figure 26. Rotation of the tower during the tilt correction on 31.10.2013.

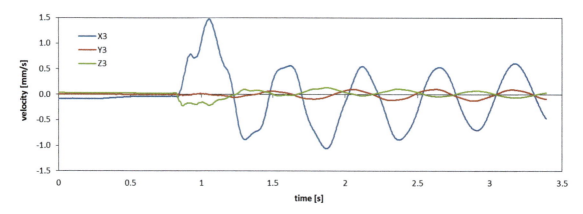

Figure 27. Free vibrations of the tower (velocity, sensor E3) during the tilt correction on 31.10.2013.

ACKNOWLEDGEMENTS

The author is grateful to the community of St. Moritz for their longstanding support and permission to publish the data. Special thanks to Ivo Sterba, formerly of IGT ETHZ, and Pietro Baracchi, formerly of the municipality of St. Moritz, for their support, friendship and many years of fruitful collaboration. Dr. Dominik Hauswirth and Balz Friedli, both of IGT, kindly helped to prepare this article.

REFERENCES

Brockmann-Jerosch, M., Heim, A. & Heim, H. (1952). Albert Heim – Leben und Forschung. Wepf & Co., Verlag, Basel.

Burland, J. B. (2008). The founders of Géotechnique. Géotechnique 58, No. 5, pp. 327–341.

Cornelius, H.P. (1912). Geologische Karte 1:25'000, Spezialkarte 115B.

Heim, A. (1932). Bergsturz und Menschenleben. Ausgabe 20 von Beiblatt zur Vierteljahrsschrift der Naturforschenden Gesellschaft in Zürich.

Haefeli, R. (1944). Zur Erd- und Kriechdruck-Theorie. Schweizerische Bauzeitung, Band 124, pp. 256–260.

Haefeli, R. (1974). Der schiefe Turm von St. Moritz im Vergleich zum schiefen Turm von Pisa. Schweizerische Bauzeitung Band 92, Heft 16, pp. 381–388.

Maillart, R. (1931). Die Erhaltung des schiefen Turmes in St. Moritz. Schweizerische Bauzeitung, Band 97/98, Heft 3, pp. 29–31.

Margadant, S. & Maier, M. (1993). St. Moritz – Streiflichter auf eine aussergewöhnliche Entwicklung. Gemeinde St. Moritz, Kur- und Verkehrsverein St. Moritz. Verlag Walter Gammeter, St. Moritz.

Marti, P. (1996). Robert Maillart – Betonvirtuose, Gesellschaft für Ingenieurbaukunst, Band 1. Vdf Hochschulverlag, Zürich.

Mueller, E.R. & Messina, G. (1992). Geotechnische Gutachten, Bericht Nr. 2570-1, Rutschung Sass Runzoel-Brattas, St. Moritz. Buechi und Mueller AG, Chur (unpublished).

Oberender, P. & Puzrin, A.M. (2016). Observation-guided constitutive modelling for creeping landslides. Géotechnique 66, No. 3, pp. 232–247.

Puzrin, A.M. & Schmid, A. (2011). Progressive failure of a constrained creeping landslide. Proceedings of the Royal Society. Series A, mathematical, physical and engineering sciences, Vol. 467, pp. 2444–2461.

Schluechter, Ch., Lang, H.J. & Wullimann, R. (1982). Brattas-Hang St. Moritz. Geotechnische Zonierung – Radiometrische Altersbestimmung. Bericht IGB, 3922/3, ETH Zuerich.

Schluechter, Ch. (1988). Instabilities in the area of St. Moritz, Switzerland. Proceedings 5th Int. Symposium on Landslides, Lausanne.

Schluechter, Ch. & Helfer, M. (1991). Der Brattas-Hang bei St. Moritz (Geologie und Morphodynamik): Eine Pilotstudie zur Verminderung von Naturkatastrophen, Ingenieurgeologie ETH Zuerich.

Schwager, M. & Puzrin, A.M. (2014). Inclinodeformometer pressure measurements in creeping landslides. Géotechnique 64, No. 6, pp. 447–462.

Sterba, I., Lang, H.-J. & Amann, P. (2000). The Brattas Lanslide in St. Moritz. Proceedings of GeoEng, Melbourne.

Tschudi, D. & Angst, R. (1999). Diplomvermessungskurs Samedan 1998, Rutschhang Brattas, St. Moritz. ETH Zuerich, Institut fuer Geodaesie und Photgrammetrie, Bericht 283, Februar 1999.

Wullimann, R. & Schneller, F. (1989). Der schiefe Turm von St. Moritz. Schweizer Ingenieur und Architekt, Band 107, Heft 35, pp. 901–906.

The towers of St. Stephen's Cathedral, Vienna

H. Brandl
Institute of Geotechnics, Vienna University of Technology, Vienna, Austria

ABSTRACT: St. Stephen's Cathedral (Stephansdom) is the dominating symbol of Vienna, constructed between 1137 and 1147. It has four towers, and the 136.7 m high main tower (South Tower) was for many years the highest tower in Europe or worldwide, respectively. This chapter gives some historical information, geotechnical details, and describes the protective measures during the immediately adjacent Vienna Metro construction including a four-storey underground station connecting the Underground lines U1 and U3.

1 INTRODUCTION

St. Stephen's Cathedral is not only the symbol of Vienna but also the most valuable building in Austria. It represents nearly nine centuries of architectural history. Fires, wars, restoration and redesign measures (e.g. from Romanesque to Gothic style) changed extent and appearance of the Cathedral over the centuries.

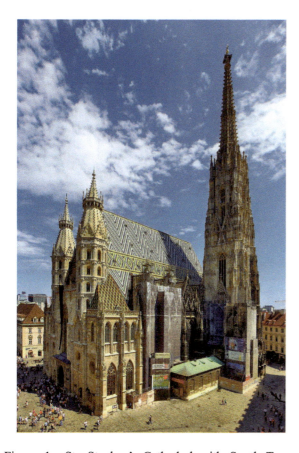

Figure 1. St. Stephen's Cathedral with South Tower (136.7 m high) and West Towers (© Bwag/Wikimedia). Local facade curtains illustrate permanent maintenance work.

Figure 2. Front view of St. Stephen's Cathedral with main entrance ("Giant's Door") and West Towers (Dombauhütte St. Stephan). Partial North view see Fig. 30a.

Figure 1 gives a view of the present building, seen from the South-West and Fig. 2 from the North-West. The construction of its most prominent feature, the "South Tower" started in 1359 and was completed in 1433; but residual works lasted until 1511. This 136.7 m high tower was not only for a long time the highest tower worldwide (at least in Europe) but had also a significant symbolic character during the permanent military attacks from the East, especially between the early 16th and late 17th century. The lookout 75 m above street level was once used as a fire warden's station and as the main observation and command post for the defence of the then walled city.

The catacombs underneath the Cathedral contain the mausoleum of the bishops, tombs and urns of the Habsburg family and bones of about 10,000 Viennese citizens.

The building is mainly founded on collapsible loess, exhibiting rather high base pressures that today would not be accepted for new buildings under such conditions. Consequently, the construction of a large Metro station (junction of Undergrounds U1 and U3 – see Fig. 28) in the heterogeneous multi-layered soil beneath the St. Stephen's Square had to prevent any geotechnical risk for this outstanding Austrian symbol.

2 HISTORICAL OVERVIEW

St. Stephen's Cathedral is located close to the ancient Roman military camp Vindobona and was erected in several construction phases (Fig. 3). The church was orientated to the sunrise on St. Stephen's feast day on 26th December, as the position stood in the year when construction began. It belongs to top ranking World Cultural Heritages.

1137–1147	Romanesque building ("Ecclesia Wiennensis")
1193	After destruction by fire new building was erected with nave and two aisles and western towers
1230–1258	Extension, construction of Romanesque "Giant's Door" (main entrance)
1304–1340	First Gothic extension (Albertine choir)

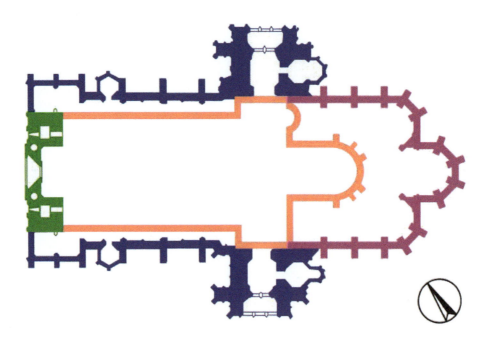

Figure 3. Ground plan of St. Stephen's Cathedral in Vienna with the different construction phases (partly on graveyards of 4th, 9/10th century), without Baroque additions (Wikipedia):
Green: Romanesque towers and "Giant's Door" from the burned first church 1137.
Pink: Romanesque second church (1263). Red: Gothic Albertine Choir (1340)
Blue: Duke Rudolf IV additions (1359), which removed the second church.

1359–1433	Construction of new Gothic nave with two aisles. Foundation stone for the South Tower (construction completed in 1511). This 136.7 m high tower has dominated Vienna's skyline for centuries.
1450	Laying the corner stone for the North Tower that was originally intended to mirror the South Tower. Construction (Fig. 4) was stopped in 1511. The total height of the unfinished Tower is 68 m.
1469	St. Stephen's Cathedral becomes mother church of canonically established Diocese of Vienna.
1522	North Tower is capped with a Renaissance cover (Fig. 38).
1529, 1683	Comprehensive repair and restoration works after Turkish sieges (Fig. 5).
1590	Strong earthquake caused severe damages, especially at the South Tower (Fig. 25). Figure 6 shows the repaired Cathedral in 1609 illustrating its dominating feature. In 1640 the Baroque influence starts with the high altar.
1783	Official closing of catacombs, but urns of the Habsburgs buried in the Imperial Burial Vault until 19th century.

Between 1839–1842 and 1861–1864 the top 17 m and 40 m, respectively of the South Tower were replaced due to an increasing inclination to the North and severe damages.

1890	A design to heighten the North Tower similar to the South Tower was discussed (Fig. 7) but never executed. (Mainly for financial reasons, but probably also because of the risk of different settlements.)

The most catastrophic event in the history of St. Stephen's Cathedral was World War II – bombing by the US Air Force on 12th March 1945 and mainly the huge fire on 11th April 1945, when the building experienced nearly complete destruction: collapse of vaulting and arches of nave and right aisle (Figs. 8, 9). Moreover, nearly all historical stained glass windows and the entire Gothic choir stall, the bells, etc. were destroyed.

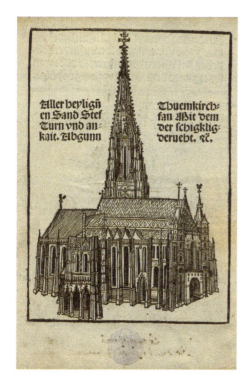

Figure 4. Early construction phase of the North Tower of St. Stephen's Cathedral. Started in 1467 but was stopped in 1511 (Dombauhütte).

Figure 5. Panoramic view of Vienna during the first Turkish siege under Sultan Süleyman, 1529, Colour lithograph, 1851 by Albert Camesina, after a coloured woodcut, 1530, by Niklas Meldemann.

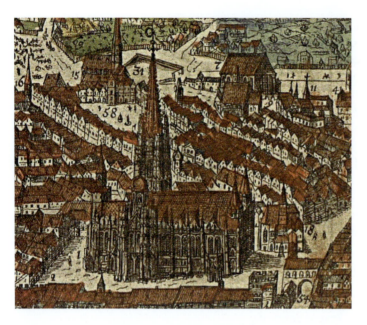

Figure 6. St. Stephen's Cathedral, a dominant feature of the historical Vienna skyline: Jacob Hoefnagel 1609 (Wikipedia). Figure 4 and 7 are from Archiv der Dombauhütte St. Stephan.

Figure 7. Design view of St. Stephen's Cathedral if the North Tower would have been completed as twin structure to the existing South Tower. Project of Friedrich von Schmidt, 1890.

Figure 8. St. Stephen's Cathedral burning in World War II from 9th to 12th April 1945 causing severe damage (Archiv der Dombauhütte St. Stephan).

Figure 9. The ruins of St. Stephen's Cathedral at the end of World War II (1945). View from the collapsed choir to the South Tower (Archiv der Dombauhütte St. Stephan).

The structural reconstruction of the Cathedral was performed between 1945 and 1952, but additional restoration work lasted until 1965. The official re-opening was celebrated on April, 26th, 1952, and in 1957 the re-founded Boomer Bell ("Pummerin") rang again in the North Tower.

With a weight of 21.4 tons it is Austria's largest bell. For safety reasons (vibrations) it is rung only on special occasions.

A building like St. Stephen's Cathedral requires permanent maintenance and local repair. Accordingly, it is hardly possible to take photographs without scaffolds or "curtains" (Fig. 1). All stonemason works have to consider the special properties of the stone materials used during the centuries, mainly lime-sandstone. The different technologies of former stone tooling, erosion resistance against weather attack, air pollution, etc. represent additional challenges of restoration, maintenance and preservation.

3 GROUND PROPERTIES, SOIL PARAMETERS

The design of twin tunnels and the Metro station immediately in front of the Cathedral required comprehensive ground investigations (boreholes with soil sampling, exploration pits, ground piezometers).

The ground below and around St. Stephen's Cathedral consists of the following main layers of locally varying thickness – e.g. Fig. 29. (The interfaces are not even but exhibit irregular profiles, e.g. ±1.5 m for the tertiary surface):

- Building and cultural rubbish and other man-made fills, locally down to 10 m with 2–2.5 m cemetery fills (soil with bones)
- Collapsible loess with small calcareous tubes (pseudomycel); locally also weathered loess-loam
- Sandy gravels of different sources (platy and rounded) with interlayers of sand and silt
- Sands
- Tertiary sediments (from silty sand to fat clay; locally with shells and cemented inclusions), belonging to the Miocene age. In K. Terzaghi's publications and up to now they are called "Wiener Tegel".

The first groundwater level was found in the quaternary sediments (about 14 m below the surface). Second and third groundwater levels (confined aquifers) were observed in sandy interlayers within the silty-clayey tertiary sediment (about 24.5 and 31.5 m below surface).

Fig. 10 shows some selected grain size curves. However, locally the heterogeneous ground exhibited a wide scatter: fat clays, uniform sands (running sands), single grain and intermittent sandy gravels, and boulders (d > 250 mm) at the interface between quaternary and tertiary sediments. The uniformity coefficient of uniform sands dropped locally to $d_{60}/d_{10} = 1.8$.

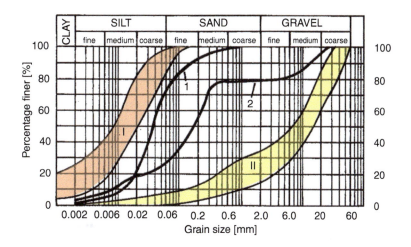

Figure 10. Some grain size distributions from a borehole close to the West front of St. Stephen's Cathedral. (Döllerl et al., 1976).
I = Tertiary sediments ("Wiener Tegel")   II = Quaternary sediments
1 = Collapsible loess   2 = Intermittent sand

The loess exhibited a rather high friction angle ($\varphi = 28° - 33°$), but a significant tendency to collapse (Fig. 11). The mean value of dry unit weight was about $\gamma_d = 17\,kN/m^3$, the oedometric modulus varied between about $E_s = 15$ to $25\,MN/m^2$ (with lower singular values).

The porosity of the sandy gravels was determined between $n = 0.2$ to 0.4. The scattering unit weight of solid particles between $\gamma_s = 26.7$ und $27.5\,kN/m^2$ could be explained by differences in the mineralogical contents.

The properties of the tertiary sediments were of greatest importance with regard to the interaction between St. Stephen's Cathedral and the different construction phases of the immediately adjacent Metro tunnels and multi-storey station. According to the wide scatter of grain size distribution also the other soil parameters exhibited a wide variety e.g. the dominating silty layers (rounded values):

Natural water content	$w = 15$–30%
Liquid limit	$w_L = 30$–55%
Plasticity limit	$w_P = 10$–30%
Plasticity index	$I_P = 10$–40%
Porosity	$n = 0.35$–0.42
Dry unit weight	$\gamma_d = 16$–$18\,kN/m^2$
Friction angle	$\varphi = 20°$–$35°$
Residual friction angle	$\varphi_r = 10°$–$30°$
Unconfined compression strength	$q_u = 0.1$–$0.9\,MN/m^2$
Oedometric modulus	$E_{s1} = 20$–$50\,MN/m^2$ (first loading)
	$E_{s2} = 40$–$80\,MN/m^2$ (second loading)
Hydraulic conductivity (laboratory)	$k = 10^{-9}$–$10^{-11}\,m/s$ for clayey silts
In-situ permeability is clearly higher:	$k = 10^{-7}$–$10^{-9}\,m/s$ in (clayey) silt, up to $10^{-4}\,m/s$ in sand.

The shear parameters were determined on undisturbed and reconstituted soil samples in a box shear apparatus which enables an alternating shearing in both directions, consequently large shear displacements. This

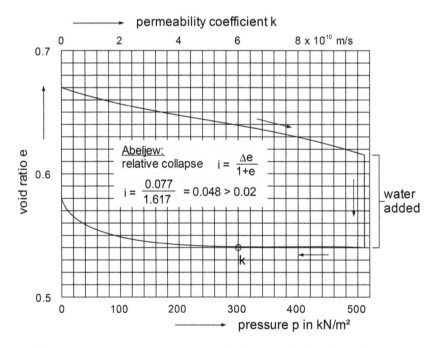

Figure 11. Pressure-void ratio diagram (from oedometer tests) of collapsible loess in the foundation zone of St. Stephen's Cathedral (example from exploration pit 1367). (Döllerl *et al.*, 1976).

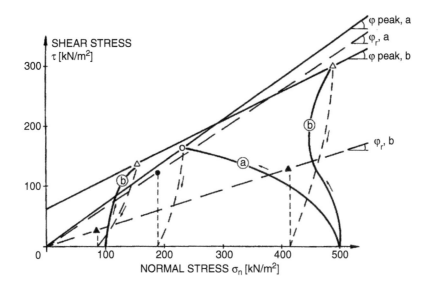

Figure 12. Vienna Shear Test: Examples of typical stress paths at constant void ratio for determining the peak shear strength, and residual shear strength after large displacements (at $\sigma_n = $ constant). (Brandl, 1993).
● ○ = normally consolidated sand (a)   △▲ = overconsolidated clay (b) (stress paths of grouted soil or cemented sand roughly similar).

"polishing" of the sample provides reliable results for the residual shear strength parameters, and the stress conditions are clearer than in case of an annular shear apparatus.

Figure 12 demonstrates the test procedure of the "Vienna Shear Test" which has proven suitable since the 1960s (Borowicka, 1970; Brandl, 1997). The test is performed with a constant void ratio to achieve the peak value of internal friction. This requires a continuous change of the effective vertical stress (σ_n, σ_v) and provides a stress path separating dilatation (left) from compression (right) which is another important soil characteristic. The tests are carried out on consolidated samples and slowly enough to ensure that the specimen remains in a drained condition. Commonly, this first shearing lasts at least 7 hours. Thus, secondary excess pore water pressures are definitely avoided.

After reaching the "intact failure" the sample is partially unloaded: $\tau \rightarrow 0$ under the condition e = const. of a constant remaining void ratio (hence also "e = const.-test"). Then shearing is performed in the opposite direction, but now as quick test with $\sigma_v = $ const. and e $\neq$ const. This alternating shearing is continued until a minimum limit value of shear strength is reached. This "polishing" gives evidence on the post-ruptures behaviour and provides the residual shear strength φ_r as a useful lower-bond or limit value (Fig. 12). Sometimes the residual shear strength may decrease significantly (down to $\varphi_r = 4°$), mainly in soils, exhibiting shiny slickenside. The limit value φ_r is of special importance for tunnelling and deep construction pits in (overconsolidated) clay or silt, and in tectonically disturbed fine-grained sediments as given in Vienna.

In a soil with stress paths according to (b) in Fig. 12 smaller deformations will be observed during tunnelling and pit excavation than in case of (a). Moreover, the dilatation after exceeding the peak value plays an important role. Its magnitude depends very strongly on the density, with denser soil expanding more rapidly. The influence of the dilatation rate on the deformation of a tunnel lining is such, that much larger movements are observed for a more dilatant soil.

Figure 13 shows typical compression curves obtained from oedometer tests on overconsolidated fine grained sediments of the "Wiener Tegel". Swelling could be widely neglected below and aside St. Stephen's Cathedral, but preconsolidation or overconsolidation, respectively had to be considered because it influenced the lateral earth pressure on tunnels, shafts and Metro station.

Therefore, oedometer tests on overconsolidated soils were not only performed in a standard manner. Experience and back analyses from numerous site measurements have disclosed that the stiffness, hence the compression modulus of overconsolidated clay, is actually higher than derived from a routine oedometer test. According to Fig. 14 the load should be increased incrementally up to the natural overconsolidation pressure, thus providing the (usual) modulus of first loading E_{s1}. After unloading, the reloading curve with a modulus

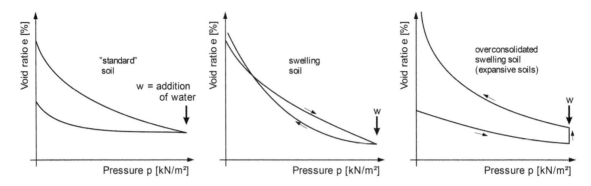

Figure 13. Typical compression curves from oedometer tests on soil samples around St. Stephen's Cathedral.

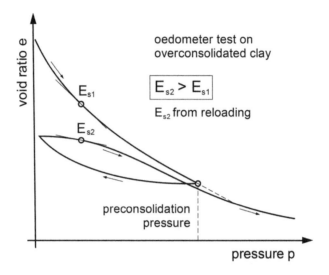

Figure 14. Oedometer test (e-p-curve) for overconsolidated tertiary sediments below St. Stephen's Cathedral – schematical: Modulus E_{s2} from re-loading curve is significantly higher than conventional E_{s1} from first-loading curve.

E_{s2} is obtained. This testing procedure idealizes the geological prehistory of the soil with a practicable accuracy. The modulus E_{s2} is typically greater than the E_{s1}-value and had proved suitable for the design of tunnel linings and for settlement prognoses.

The overconsolidation or preconsolidation pressure of the soil was determined by drawing the e-p-curve in a semi-logarithmic scale (Fig. 15). At the preconsolidation pressure the curve rather abruptly bends downward, turning into a straight line that represents the stress-settlement property for normal consolidation. There are different possibilities of interpretation, but the indicated poc determination has proved to be in good accordance with the comprehensive geological investigations of the Vienna's ground conditions.

Figure 15 also demonstrates the anisotropic behaviour of an overconsolidated tertiary clay that had undergone severe tectonic stresses. Consequently, the preconsolidation pressures in the vertical and horizontal direction are not identical. Such a difference can be observed also for the moduli E_s. Therefore, oedometer tests were performed on undisturbed samples with different loading directions. The sophisticated tunnel design was based on E_v- and E_h moduli of the surrounding soil, thus considering a different behaviour of the vertical and horizontal ground reaction.

From Figure 15 it also can be clearly seen that the amount of swelling is strongly influenced by the angle between fissures and/or layers and the direction of loading/unloading p. Horizontally layered soil exhibits a greater swelling potential.

Experience has shown that there is no need to determine the preconsolidation or overconsolidation ratio (OCR) for all samples of a site investigation programme. Usually, the magnitude is roughly the same within a site, and some exploration tests are sufficient. Around St. Stephen's Cathedral the overconsolidation pressure

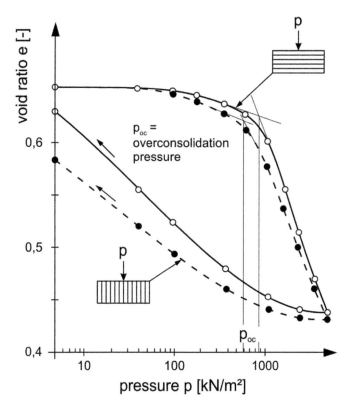

Figure 15. Determination of overconsolidation pressure by oedometer tests. Pressure p in logarithmic scale. Example for tertiary sediments in Vienna illustrating different vertical and horizontal behaviour.

lies between $p_{oc} = 1.5$ to $2\,MN/m^2$. Samples near the tertiary-quaternary interface frequently show smaller values due to softening and weathering in geological times.

The coefficients for the earth pressure at rest of the tertiary sediments were derived from triaxial tests on undisturbed samples. The results were mostly somewhat higher than calculated from Jaky's formula $K_0 = 1 - \sin\varphi$, adapted from originally $K_0 = 0.9(1 - \sin\varphi)$, where $\varphi =$ friction angle. The difference can be explained by ground-inherent residual stresses from overconsolidation and tectonic constraints. As design values $K_0 = 0.5$ to 0.7 were assumed, depending on soil properties and ground structures.

In general, rock permeability and groundwater velocity depend on the shape, amount, distribution and interconnectivity of voids. Voids, on the other hand, depend on the depositional mechanisms of carbonate sedimentary rocks, and on various other geologic processes that affect all rocks during and after their formation (see Chapter 2). Primary porosity is the porosity formed during the formation of rock itself, such as voids between the mineral grains, or between bedding planes. It is also often called matrix porosity. Secondary porosity is created after the rock formation, mainly due to tectonic forces (faulting and folding) which create micro and macro fissures, fractures, faults and fault zones in the brittle rock such as limestone. Sedimentary carbonate rocks may become cavernous (karstic) as a result of the removal of part of its substance through the solvent action of percolating water. Although solution channels and fractures may be large and of great practical importance, they are rarely abundant enough to give an otherwise dense rock a high porosity.

4 FOUNDATION OF THE TOWERS

4.1 Type and quality

Figure 16 shows the close vicinity between St. Stephen's Cathedral and the structures for the Metro lines (Undergrounds U1 and U3) and connecting station. Due to the minimum distance of 5 m interactions had to be considered and required a detailed investigation of the Cathedral, especially of its four towers.

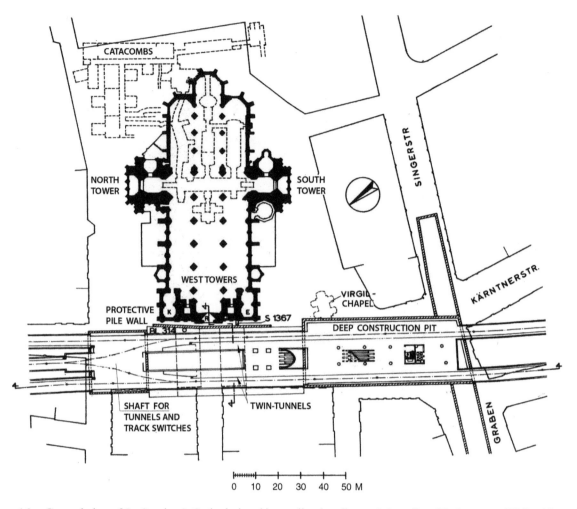

Figure 16. Ground plan of St. Stephen's Cathedral and immediately adjacent Metro line (Underground U1) with station and connection to the Underground U3. (Döllerl et al., 1976).

The majority of foundations stand on collapsible loess, only the south-west corner rests on sandy gravel. However, neither the subsurface church of the Cathedral nor the catacombs reach near to the western front with the main entrance and the adjacent Metro structures.

Detailed investigations of the foundations (Kieslinger 1949, 1971) showed that there is an extremely different quality regarding stone material, tooling and stonemason school due to the different construction periods. Frequently thick stone plates and small stone pieces up to large blocks of nearly 1 m were used rather varyingly as rubble masonry (Fig. 17). The Romanesque central part with the "Giant's Door" and the western towers is founded on calcareous sandstone, limestone, quartzitic sandstone from the near flysch zone and on numerous rests of ancient Roman fortifications. The younger footings have a more or less similar composition, whereby Gothic and Romanesque parts were separated by a joint (Fig. 18).

The cementation of the foundation masonry was also found to be rather heterogeneous: locally mortar of high quality, but frequently with insufficient or leached out mortar or even without mortar.

A cross section illustrates the different construction phases and subsurface structures (Fig. 19). The columns are founded on collapsible loess. Figure 20 is a cross section through the South-West Tower near the main entrance ("Giant's Door") of the Cathedral. This tower is founded on silty sandy gravel, or at least in the interface zone between loess and gravel.

Experience has shown, that most buildings where heavy bombing occurred nearby or destroyed parts of these buildings during World War II exhibit residual stress constraints in their structure. Consequently, they are rather sensitive towards stress rearrangements caused by adjacent construction work. This had to be considered also for St. Stephen's Cathedral that had undergone severe damages in the year 1945 (Figs. 8, 9).

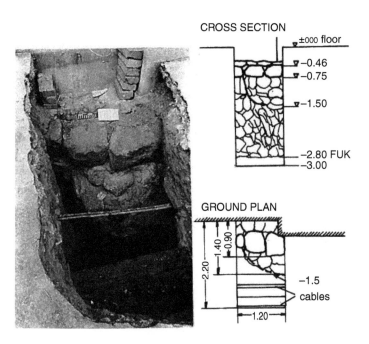

Figure 17. Typical exploration pit to investigate the foundation of St. Stephen's Cathedral. (Kieslinger, 1971).

Figure 18. Exploration pit showing the direct contact between the Romanesque foundation of the North-West Tower and the Gothic foundation of the adjacent Cross Chapel. Separated by a joint. (Kieslinger, 1971).

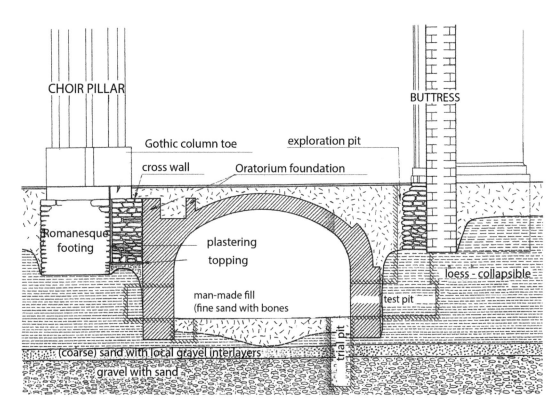

Figure 19. Partial cross section through St. Stephan's Cathedral showing the locally different foundation conditions and soil properties (Kieslinger, 1971). Soil layers usually not horizontal.

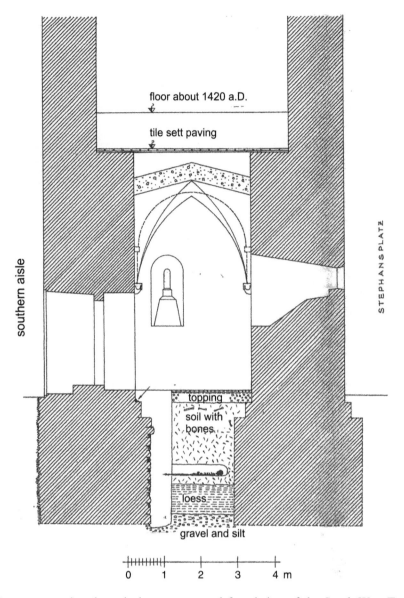

Figure 20. East-West cross section through the toe zone and foundation of the South-West Tower of St. Stephen's Cathedral (Kieslinger, 1971).

4.2 *Soil pressures (base pressures of foundations)*

The soil pressures below the foundation of the structural main parts of St. Stephen's Cathedral are very high, reaching 800 kN/m² for foundation depths between 2.2 to 3.3 m. This clearly exceeds the "critical edge pressure" that indicates the beginning of a theoretically plastified zone below a foundation. Fig. 21 illustrates this gradual ground plastification according to O.K. Fröhlich's theory (1936), that provides quite reasonable results, as experience has confirmed.

For the layer immediately below the foundation the critical edges pressure q_{cr} is given by

$$q_{cr} = \frac{\pi(p_k + \gamma \cdot t)}{\cot \varphi - \left(\frac{\pi}{2} - \widehat{\varphi}\right)} \qquad (1)$$

with t = foundation depth, γ = unit weight above foundation base,
$p_k = c.\cot \varphi$, c = cohesion, φ = friction angle below foundation base

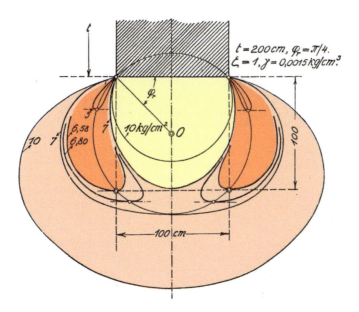

Figure 21. Progressive expansion of plastified ("flowing") zones below a flat foundation with increasing soil pressure emanating from the outer edges. Example for foundation depth $t = 2$ m of a strip foundation (after O.K. Fröhlich, 1934 – coloured).

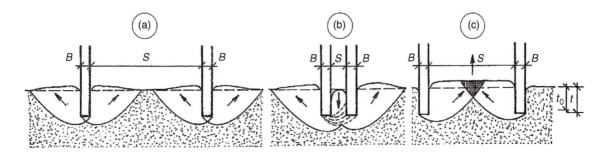

Figure 22. Model tests to investigate the interaction of close foundations on their bearing capacity (Myslivec, 1968).

Log-term experience has shown that q_{cr} may be exceeded by 20% without relevant soil plastification and practical problems (i.e. larger settlements). Though this would result in about $q_{cr} = 400 \text{ kN/m}^2$ (for $c = 0 \text{ kN/m}^2$), it is still only 50% of the existing maximum ground pressure of 800 kN/m². However, the stress indicating the beginning of progressive failure lies about $\sigma_{BB} = 1200 \text{ kN/m}^2$, if cohesion is neglected. (O.K. Fröhlich's term σ_{BB} comes from the German „Bruchbeginn – Spannung"). Considering a reasonable amount of cohesion would increase O.K. Fröhlich's q_{cr} and σ_{BB} by about $\sigma_{BB} = 200 \text{ kN/m}^2$. Moreover, σ_{BB} does not describe collapse but approaching ultimate limit state.

Settlements of buildings are caused not only by compression but also by shearing of the underlying soil. Therefore, strong plastification trigger a gradual reduction of the peak shear resistance towards residual values, thus decreasing stability against ground failure and increasing settlements. From this point of view O.K. Fröhlich's q_{cr} is a valuable tool for risk assessment of foundations on soil with a low residual shear strength. It provides a very cautious border value.

Historical foundations were sometimes not monolithic blocks but contain a central core filled with rubbish or soil being enclosed by stone masonry (e.g. Fig. 20). Moreover, the foundations may be closely spaced, thus, interacting in their bearing-settlement behaviour. In case of St. Stephen's Cathedral the close foundation parts of the South-West tower, etc. interact in a positive way. Investigations of A. Myslivec (1968) showed that the bearing capacity of close foundations at first increases for narrow spacings before decreasing at wider spacings (Figs. 22, 23). The maximal loaded foundations exhibit a S/B – value of about 0.5 to 0.8, thus having a higher bearing capacity than single foundations, because they act more or less as one unit together with the

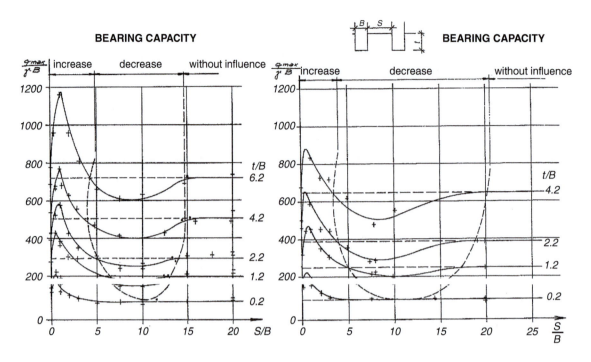

Figure 23. Two results to Fig. 22 illustrating increase of bearing capacity for narrow spacing, followed by decrease and finally no interaction (Myslivec, 1968).

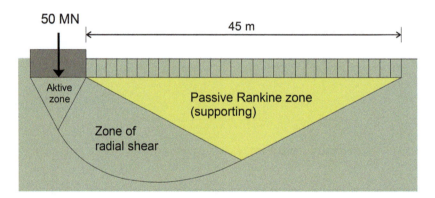

Figure 24. Ground failure calculation according to Eurocode 7 for South-East Tower of St. Stephen's Cathedral. Example of parametric studies showing the far reaching passive Rankine zones. Only drawn in direction of St. Stephen's Square south of the Cathedral.

enclosed soil. Hence, this quasi-composite system can be considered a "box-foundation", where the soil core contributes to the bearing capacity (Brandl, 2010).

For monolithic foundations and partly for box-shaped foundations conventional theories based on global safety factors, and Eurocode 7 assumptions based on partial safety factors facilitate reliable risk assessment of historical foundations, but should be combined with parametric studies (varying ground parameters, foundation dimensions, etc.). Moreover, such calculations indicate the far-reaching interaction between foundations and ground as shown for the South-West Tower of St. Stephen's Cathedral (Fig. 24). There the passive Rankine zones reach about 45 m aside of the foundation. Disturbing this lateral support by excavations without geotechnical plans and stabilizing or compensation measures would decrease the safety factor against ground failure. This fact is frequently not known or ignored by archaeologists, thus reducing the safety against ground failure when excavating even relatively far aside historical buildings.

Finally, the slow rate and repeated interruption of construction of the towers were certainly of great advantage for their geotechnical stability: dissipation of pore water pressures, but gradual compression and not sudden collapse of loess.

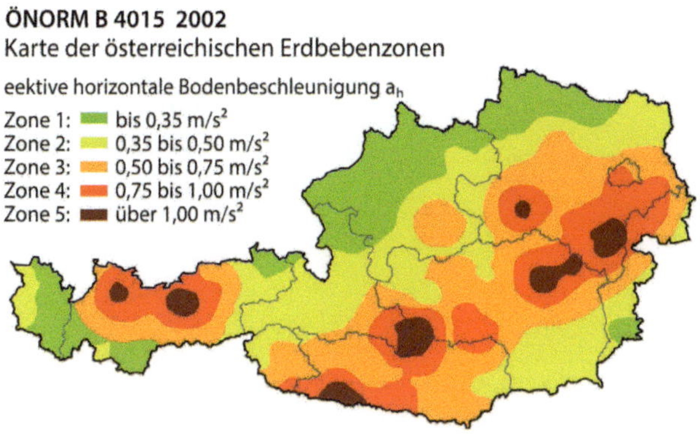

Figure 25. Seismic zones in Austria after Austrian Standard ÖNORM B4015-2002. Effective horizontal soil acceleration a_h (Zentralanstalt für Meteorologie und Geodynamik, Wien, 2002).
St. Stephen's Cathedral lies in zone 4 ($a_h = 0.75$–1.00 m/s^2).

Figure 25a. Part of a historical newspaper (Schultes, Hans A., 1590) showing the severe damages of St. Stephen's Cathedral and other buildings in Vienna during a strong earthquake (intensity 9, magnitude 6.2 in 1590). (Städtische Kunstsammlung Augsburg).

To sum up, the soil pressures below the old foundations of St. Stephen's Cathedral are clearly higher than today acceptable for new buildings (especially in collapsible loess) and according to present standards, e.g. Eurocode 7; but they have been proving sufficient bearing capacity for centuries. This is also the case for numerous historical buildings and requires their cautions treatment or strengthening if structural measures are taken within the old building or in close vicinity. Therefore, considering the "Observational Method" defined by geotechnical engineering should be obligatory.

4.3 *Earthquakes*

Vienna lies in seismic zone 4, where an effective horizontal soil acceleration of $a_n = 0.75$ to 1.00 m/s^2 has to be considered for design (Fig. 25). Eurocode 8 (EC8) has changed the seismic risk assessment for endangered buildings in Austria's urban centers significantly. The earthquake 1590 (magnitude 6.2 after Richter) caused severe damages also on St. Stephen's Cathedral (Fig. 25, 25a).

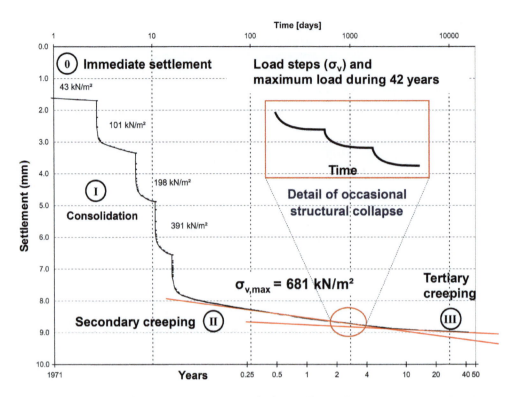

Figure 26. Results from a long-term oedometer test on soft clayey silt running 42 years (1971–2013). Settlements under increasing load steps and long-term creeping under maximum pressure (681 kN/m^2), illustrating that creeping of normally consolidated clayey soil under high loads never stops. Time in logarithmic scale (Brandl, 2013).

During earthquakes the edge pressures in the foundation bases are temporarily increased. However, towers like those of St. Stephen's Cathedral are usually not very much endangered to failure due to their low tuning (their natural frequency is well below the excitation frequency). This could be found also during the pile wall installation in front of the Cathedral on 12th April 1972. Moreover, the dominating first natural frequency (horizontal vibration) can be reduced by a vibration absorber if necessary.

4.4 *Settlements*

The different foundation depths, heterogeneous subsoil conditions and locally very high soil pressures created differential settlements of the structure up to about 0.15 m. The total settlement of the high South Tower during its construction phase is estimated to be about 0.2 to 03 m.

Due to the large foundations and high base pressure of the towers the "settlement-relevant" stress zones in the subsoil reach the tertiary sediments which exhibit a widely uneven surface (roughly ±1.5 m). The deep reaching stress distribution in the clayey-silty layers has been the reason for ongoing secondary/tertiary settlements, i.e. long-term creeping. Contrary to K. Terzaghi's opinion, creeping of (high loaded) clayey-silty soil never stops as could be proven by H. Brandl, 2013 (Fig. 26). This is due to microcrystalline rearrangements caused by ductile sliding between mineral crystals. Consequently, creeping depends on numerous factors as grain size distribution, sensitivity, soil fabrics, mineralogical composition, viscosity, permeability, etc. Fading out of creeping occurs only if the soil pressure lies clearly below a long-term overconsolidation pressure.

5 METRO DESIGN AND CONSTRUCTION – INTERACTION WITH CATHEDRAL

5.1 *Design*

The Metro construction in front of St. Stephen's Cathedral comprised not only the Underground lines U1 and U3 but also a four-storey station (Figs. 16, 28). U1 and the entire junction facilities were constructed in the 1970s, whereas residual connecting works for U3 followed between 1985 and 1990.

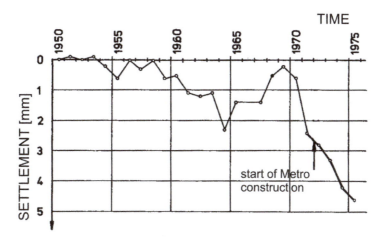

Figure 27. Settlement of the South Tower between 1950–1975. (Döllerl *et al.*, 1976).

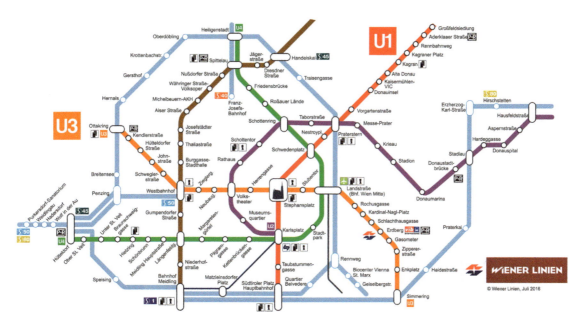

Figure 28. Partial ground plan of the net of Vienna Metro lines with Undergrounds U1 and U3 crossing at the Station "Stephansplatz" in front of St. Stephen's Cathedral. (Wiener Linien).

Several alternatives were investigated to minimize any risk for St. Stephen's Cathedral during the Metro construction. A removing of the alignment or its higher level had to be excluded because of the deep cellars of the surrounding buildings and due to operation reasons.

The final design philosophy was not to touch the Cathedral and to avoid any work within this sensitive historical building. Consequently, underpinning of the foundations with root piles or grouting of their subsoil was excluded, as well as structural measures with girder grills, etc. Underpinning with root piles (small diameter bored piles with high pressure concreting) commonly cause settlements of several millimetres during installation, and grouting involved the risk of undesirable settlements/heaves (according to then used technology). Such movements could have endangered sensitive structural elements of St. Stephen's Cathedral (e.g. Fig. 30a).

This resulted in a design separating the ground influence by Metro construction from the ground below the Cathedral. It was obtained by a protective wall of 50 inclined bored piles ($d = 0.9$ m, $\alpha = 5°$) horizontally supported by grouting the quaternary layers above the Metro (Figs. 29, 30). The grouted body served also against possible soil loss and as a tunnel roof against possible air piping during tunnel excavation under

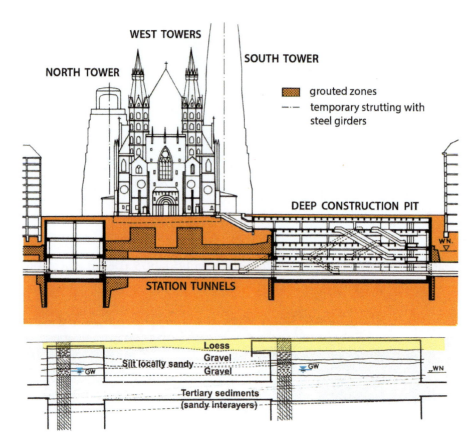

Figure 29. Longitudinal section to Fig. 16 with ground properties along U1 and construction measures (1971–1977). (Döllerl *et al.*, 1976).

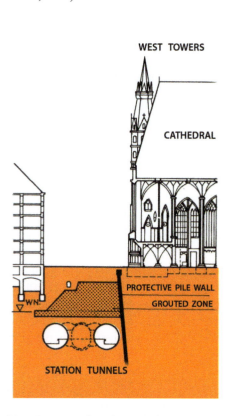

Figure 30. Cross section through the West Towers and U1 tunnels of the Metro station. (MA 38).

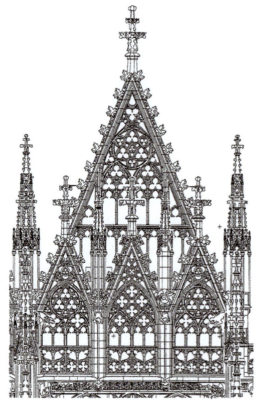

Figure 30a. Partial view of North front above Bishop's Door. (Schnabel, ÖIAZ 1993).

Figure 31. Tunnel shield (open type manual, Ø 8 m) being lowered into the staring shaft in front of St. Stephen's Cathedral. (Archiv der Dombauhütte St. Stephan).

compressed air. Another design requirement was a minimal disturbance of the aquifer and the groundwater communication, especially for the final state after construction.

Moreover, the originally designed deep construction pit in front of the Cathedral was replaced by twin tunnels (Ø = 8 m), excavated by an open type shield (manual shield) under 0.5 to 0.7 atm of compressed air (Fig. 31). The tunnel lining consists of cast-iron segments. In the starting zone for tunnel driving the tertiary sediments were also grouted. Cross connections were excavated by the shotcrete method (NATM).

5.2 *Construction*

The piles for the protective wall (depth = 32 m) were installed as intermittent elements with a free spacing of at least 0.1 to 0.3 (depth = 40 m), and those of the shaft of the track switches are tangent with local grouting (sealing) between them. The pile toes are embedded in a layer of stiff tertiary sediments. In order to minimize vibrations the entire pile excavation was performed with a hydraulic casing machinery combined with an equipment for dry rotary drilling and a Kelly rod (Fig. 32).

Slurry trench walls (thick diaphragm walls) were excluded, because many old unknown cellars and cavities could have caused numerous sudden (and excessive) losses of slurry (bentonite suspension), thus involving a high risk of trench collapse. Moreover, slurry trench excavation would have required local chisel work, especially in the boulder zone between quaternary and tertiary sediments and in cemented tertiary inclusions, thus creating critical vibrations.

During pile excavation drilling obstacles could be managed by roping down workers within the pile casing and cautiously loosening the ground with pneumatic air hammer. Moreover, the casing of the piles always had to be screwed in ahead of soil excavation in order to avoid local ground disturbance or hydraulic base failure. Consequently, there was no damage during the earthquake on 16th April 1972 (6° MS).

The large construction pit was excavated in three steps to keep the horizontal movements of the retaining structures and ground disturbance a small as possible. Finally, the pile walls of this four-storey structure were stiffened as much as possible by permanent reinforced concrete beams, using the floor structures, in order to minimize horizontal movements (Fig. 33). Temporary steel struts were installed only locally as additional

Figure 32. Dry rotary drilling equipment with telescoping Kelly bar and ahead screwing in the steel casing to minimize soil disturbance during excavation of bored tangent piles (d = 0.9 m, 40 m deep) close to St. Stephen's Cathedral in 1973.

Figure 33. View from the Cathedral's roof into the deep construction pit for the four-storey Metro station (junction of Metro lines U1 and U3 – see Fig. 28, Wiener Linien).

elements and prestressed such that beams and walls had an intensive contact from the very beginning. The vibrations during the construction period were limited by maximum acceleration of $b = 20$ cm/sec^2. For permanent operation $b = 2$ cm/sec^2 and for occasional situations 5 cm/sec^2 have been allowed.

5.3 *Construction control, monitoring*

Comprehensive monitoring was performed during the entire construction period starting with zero readings already before auxiliary works began. This comprised geodetic survey, control of vertical and horizontal movements inside and outside the Cathedral (Fig. 34), control of inclination of towers and Romanesque west front (foundation depth only 2.8 m), crack control (1/1000 mm accuracy), vibration control (acceleration, velocity, vibration path) and piezometer readings of different ground water stories, responding independently of each other to the construction measures. The measurement devices were connected with an alarm system that should alert, if limit values were exceeded. During critical construction phases readings were performed several times a day. (Today's continuous remote reading was not yet available then). A free-lance civil engineer of comprehensive experience was engaged for permanent monitoring and quality assurance. He was personally responsible for St. Stephen Cathedral's safety and serviceability.

Moreover, numerous soil samples were taken during the excavation of piles, construction pits and tunnels to compare the design soil parameters with the ground properties actually found during construction. Historical finds from Roman times and the Middle Ages could be well preserved, including the Virgil chapel which has become a part of the Metro station (Fig. 16).

During the excavation of the deep construction pit and shaft for the track switches the second groundwater level within the tertiary sediments (confined aquifer) was relieved by standpipes. Figure 35 shows the influence of geotechnical measures on the tertiary groundwater level, clearly illustrating the effects of pile excavation and sealing measures. Long-term monitoring results for the South Tower are given in Fig. 36, indicating a

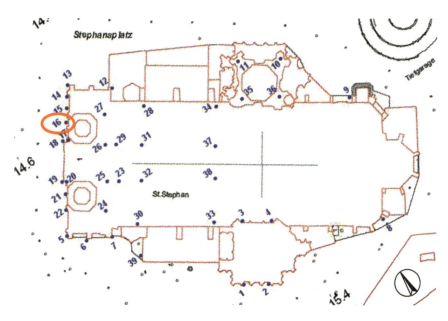

Figure 34. Ground plan with settlement markers for monitoring St. Stephen's Cathedral (Vienna Municipal Departments MA 29, MA 41).

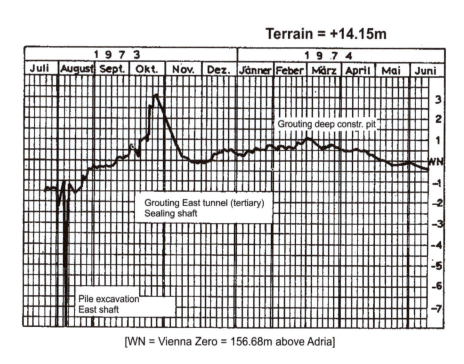

Figure 35. Temporary influence of construction measures on the tertiary groundwater level (2nd GW level, confined aquifer). (Döllerl et al., 1976).

certain communication between settlement/heave and seasonal fluctuation of the quaternary ground water level. Additionally, long-term settlements can be observed, confirming the ongoing creeping of the ground. In 1971 the auxiliary measures (relocation of pipe lines, cables, etc.) started as preparation work for the main construction stages of the Metro project. The clear influence on aquifer and on the settlement behaviour of the South Tower illustrates the sensitiveness of the geotechnical system. The settlement curve in West-East direction underlines the protective effect of the pile wall during tunnel driving and the far-reaching settlement after completion (Fig. 37).

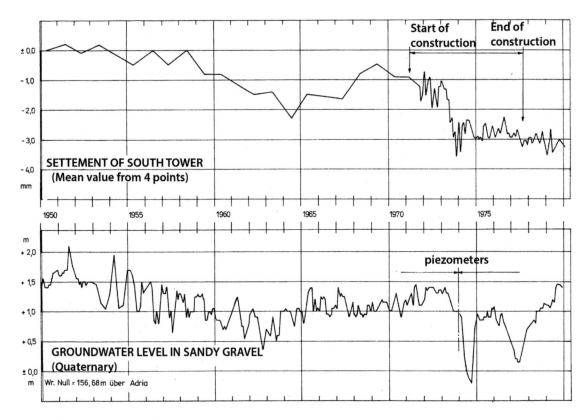

Figure 36. Time – settlement line (mean value of 4 markers) and time – groundwater level line for the South Tower of St. Stephen's Cathedral. (Hondl & Martak, 1980).

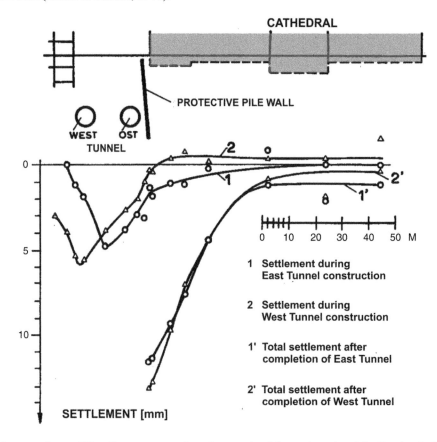

Figure 37. Settlement along a West-East cross section, 6 m north of the main axis of St. Stephen's Cathedral. (Döllerl et al., 1976).

Figure 38. Aerial view of St. Stephen's Cathedral showing all four towers and the settlement marker no. 16 on toe of North-West Tower. (Archiv der Dombauhütte St. Stephan).

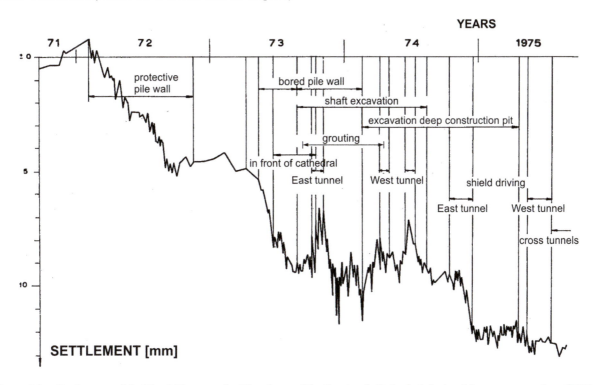

Figure 39. Settlement of the North Tower at the West front of St. Stephen's Cathedral during Metro construction. (Döllerl et al., 1976).

Comprehensive site supervision and monitoring disclosed that the allowable limit values of (differential) settlements, tower inclination and vibrations were never exceeded during the entire construction period. The data of movement marker no. 16 (Figs. 38, 39) illustrate the influence of the construction measures on the settlement of the North-West Tower at the western front (= main entrance) of the Cathedral. The installation of the bored pile wall (protective wall) contributed most because of unavoidable vibrations during pile excavation

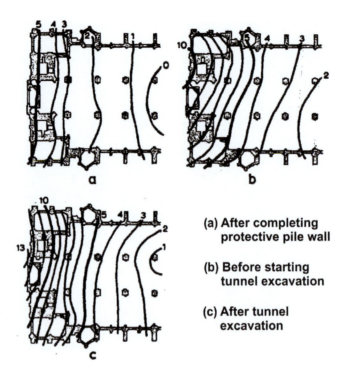

Figure 40. Settlement isohypses of St. Stephen's Cathedral during Metro construction (in mm). (Döllerl *et al.*, 1976).

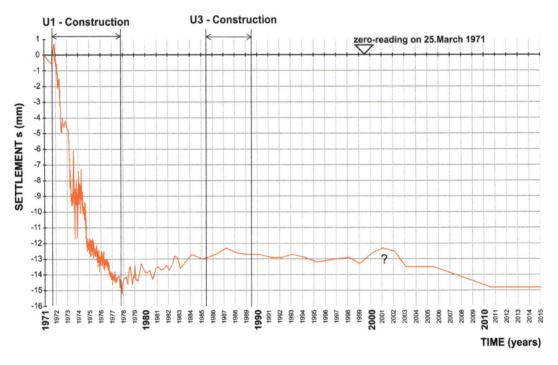

Figure 41. Time-settlement line for marker no. 16 (North-West Tower – see Figs. 34, 38) between 1971 and 2015. Construction phases of Metro lines U1 and U3 also shown. (Vienna Municipal Departments MA 29, MA 41).

and significant stress redistribution in the soil. Shield driving under compressed air, however, caused relatively small settlement. Temporary heaves were due to grouting.

The ground plan with the settlement isohypses shows rather uniform deformations with a maximum close to the North-West Tower at the western front of the Cathedral (Fig. 40). This was caused by the deep shaft excavation for the tunnel shield and the track switch, and by a thicker loess layer. The start-up phase of shield driving obviously had no relevant influence on the settlements. Figure 41 presents the time settlement curve for the North-West Tower that underwent the maximum settlement during the Metro construction. Temporary increase of groundwater level and swelling of fine-grained sediments caused temporary heave, followed by ongoing secondary settlement (long-term creeping). Reversible heave/settlement and long-term creeping will continue, but have no influence on the geotechnical stability of this tower. This conclusion can be drawn also for the other towers of St. Stephen's Cathedral.

5.4 *Long-term control*

The outstanding importance of St. Stephen's Cathedral and the interest in long-term interaction behaviour between Metro, Cathedral and groundwater led to the decision to continue monitoring. Meanwhile this has become essential according to Eurocode 7 and the "Observational Method". In detail geotechnical monitoring of St. Stephen's Cathedral comprises:

– Weekly measurements of piezometers in two aquifers (groundwater storeys).
– Geometrical levelling of all markers of the Cathedral until 1981 four times a year, then twice a year (depending on registered values).
– Convergence measurements of Metro line and tunnels in four month intervals.

Figure 42. Change of scaffolding of the South Tower during three centuries (1863, 1954, 2007) (Archiv der Dombauhütte St. Stephan).

6 SUMMARY

St. Stephen's Cathedral in Vienna represents Austria's history in stone. It is also regarded as a landmark of Austria, forms the defining urban center of Vienna and is above all a work of European rank, unique in its form. Wars and earthquakes attacked the Cathedral during centuries, and nowadays air pollution makes its maintenance demanding and costly (Fig. 42). Several extensions of the building, changes of style and other construction measures gave St. Stephen's the present feature. The construction of the immediately adjacent Metro line U1 and its four-storey station in the early 1970s were a geotechnical challenge that could be managed without any damage of the Cathedral. Continuous monitoring has been performed since the end of Metro construction in 1975, confirming that the ground engineering measures were successful and that the foundation of the Cathedral meets the high standards of Eurocode 7, based on the "Observational Method".

REFERENCES

Borowicka, H. (1970) *10 Jahre Wiener Routinescherversuch.* Mitteilungen des Institutes für Grundbau und Boden-mechanik, Technische Universität Wien, Heft 11, Wien, Juni 1970.

Brandl, H. (1997) *Residual shear behaviour and progressive failure of soil or weak rock in tunnelling.* In: Golser, J., Hinkel, W.J., Schubert, W. (ds.) World Tunnel Congress 1997. Proceedings, Vol. 1, 12–12 April 1997, Vienna. A.A.Balkema/Rotterdam, pp. 15–22.

Brandl, H. (2010) *Cyclic preloading of piles and box-shaped deep foundations.* In: International Geotechnical Conference "Geotechnical Challenges in Megacities" Moscow, 2010. Keynote Lecture. Proceedings of TCs of the International Society for Soil Mechanics and Geotechnical Engineering (Russian Member Society of ISSMGE). Moscow. Vol. 1, pp. 3–28.

Brandl, H. (2013) *Consolidation/Creeping of Soils and Pre-treated Sludge.* In: 5th Biot Conference on Poromechanics (in memory of Karl von Terzaghi, 1883–1963), Vienna 10–12 July 2013. Ch. Hellmich, B. Pichler, D. Adam (eds). American Society of Civil Engineers (ASCE), Reston, VA, USA, pp. 1346–1357.

Döllerl, A., Proksch, E., Hondl, A. (1976) *Der U-Bahn –Bau und die Maßnahmen zum Schutz des Stephansdomes.* Mitteilungen des Institutes für Grundbau und Bodenmechanik (and Proceedings VI ECSMFE in German). Technische Universität Wien, Heft 14. (Figures translated into English with some additions).

Fakultät für Architektur und Bauingenieurwesen der TH Wien (1971). *Gutachten über den Schutz des Domes von St. Stephan während des Baus der U-Bahn* (not published).

Fröhlich, O.K. (1934) *Druckverteilung im Baugrunde.* Springer, Wien.

Fross, M. (1973) *Untersuchungen über die Zusammendrückbarkeit vorbelasteter toniger Böden des Wiener Beckens.* Mitteilungen des Institutes für Grundbau und Bodenmechanik, Technische Universität Wien, Jänner 1973, Heft 12.

Fross, M. (2015) *Über Bauwerkssetzungen.* VÖBU (ed.), Wien. ISBN 978-3-902 450-05-05

Hondl, A., Martak, L. (1980) *Sicherungsmaßnahmen für den Stephansdom während und nach dem U-Bahnbau.* Der Dom – Mitteilungsblatt des Wiener Domerhaltungsvereins. Folge 1/1980, pp. 8–11.

Kieslinger, A. (1949) *Die Steine von St. Stephan.* Herold-Verlag, Wien.

Kieslinger, A. (1971) *Untersuchung des Baugrundes vor dem Stephansdom in Wien.* (not published expertise).

Myslivec, A. (1968) *Einfluß der benachbarten Streifenfundamente auf ihre Tragfähigkeit.* 2nd Danube-European Conference on Geotechnical Engineering , Vienna 8–10 May 1968, Vol. 2, Eigenverlag ÖIAV (Österreichischer Ingenideur- und Architekten-Verein), pp. 42–50.

Vienna Municipal Department MA 38 (1985) *Die 1. und 2. Ausbauphase der Wiener U-Bahn (1969–1993).* Compress Publisher, Vienna 1985.

Wiener Linien (2006) Die Linie U1; *Geschichte – Technik – Zukunft.* Vienna (ISBN-13: 978-3-200-006856-2).

Zöhrer, K. (1993) *Der Dom zu St. Stephan.* Österreichische Ingenieur- und Architekten-Zeitschrift (ÖIAZ), Vol. 12, Dec. 1993. Vienna, pp. 454–478.

Geotechnics and Heritage: Historic Towers – Lancellotta, Flora & Viggiani
© 2018 Taylor & Francis Group, London, ISBN 978-1-138-03272-9

The tower of the Admiralty in Saint Petersburg

V. Ulitsky, A. Shashkin, C. Shashkin & M. Lisyuk
Georeconstruction Engineering Co., Saint Petersburg, Russia

ABSTRACT: This contribution deals with the historical Admiralty Tower and building in central Saint Petersburg. The tower was erected in 1734 and expanded in 1823. The authors inspected the monument, analyzed the structures, soil and foundations, and performed soil-structure interaction calculation of the monument.

1 INTRODUCTION

Historical analysis of a geotechnical problem is very important for projects dealing with preservation of historical monuments and their foundations (Burghignoli *et al.*, 2007; Ulitsky, 2003).

Such analysis should include the following steps:

- Analysis of the actual stress-strain conditions of subsoil of preserved buildings, and, if necessary, of adjacent buildings;
- Estimation of the influence of present vibration background on settlement development;
- Estimation of ongoing settlements of buildings (under their own weight and outside factors), that is defined through calculations or observations of the location of geodetic marks and gauges;
- Estimation of an allowable additional settlement of the existing buildings during reconstruction works or new development.

For important projects it is also necessary to perform the following works:

- Historical analysis of foundation behaviour of preserved/reconstructed buildings and buildings adjacent to an object under reconstruction or of new development together with substructure behaviour of the existing buildings;
- Calculations of the total assumed deformations and percentage of different causes of settlement development of the existing buildings.

This contribution deals with the Admiralty building in central St. Petersburg (Fig. 1).

The building has well pronounced cracks. The purpose of the historical analysis in this case was to find out the reason of these cracks' development to make a decision about further strengthening or preservation of the monument. In 2003 during preparatory works for celebration the 300th anniversary of St. Petersburg the restorers found well pronounced cracks in the Admiralty façade. St. Petersburg Committee for Protection of Historical Monuments commissioned the authors of the present contribution with a task to investigate the state of the Admiralty building and, if necessary, to elaborate measures to reinforce its bearing structures.

2 THE HISTORICAL BACKGROUND

The main tower of the Admiralty was erected in 1734 by the architect I.K. Korobov replacing the old wooden building (Figs. 1–3). In 1811–1823 the building was reconstructed according to the project of the architect A.D. Zakharov. During the reconstruction, the height and sizes of the tower in plan were increased (Fig. 4), the internal rearrangement was made, there were installed two internal staircases and the height of the entry arch was augmented.

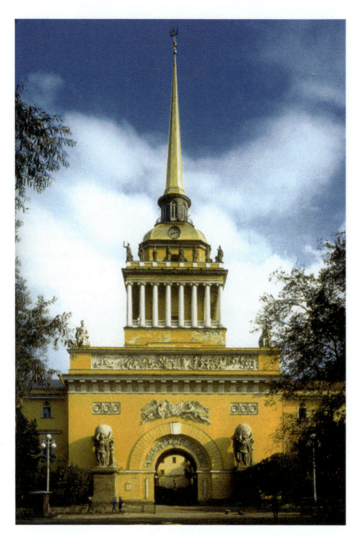

Figure 1. The Admiralty building in central Saint Petersburg.

The new walls of the main and backyard façades rested upon the new separate foundations immediately adjacent to the existing foundations. It resulted in differential settlements ongoing until now.

According to the existing data of the archive of the Committee for Protection of Historical Monuments, during the war of 1941–1945 the balustrade was partly destroyed as a result of artillery shells hitting. In 1945 and in 1951 the tower was restored by the institute "Lenproject".

In 1995 the tower was examined by the firm "Dionis" without test pits. The report stated the absence of cracks movement as the gauges of the main cracks remained undisturbed since 1971. The archive report declared no need to reinforce foundations and soils by grouting. It shows that the cracks had been identified in 1845 and remained in stabilized state, the maximum brickwork crack width being 30 mm.

In 2002–2003 the tower was investigated by two geotechnical companies. During this investigation there were dug four test pits and drilled seven boreholes. The investigation showed that the foundations settlements had not stabilized and the tower wall cracks opening had been ongoing.

3 GEOTECHNICAL CONDITIONS OF THE SITE

The analysis of geotechnical conditions was conducted based on the archive data of the Trust for Geodetic Surveys and Engineering Investigations as well as the Contractor.

Figure 2. The Admiralty building façade. The architect I.K. Korobov.

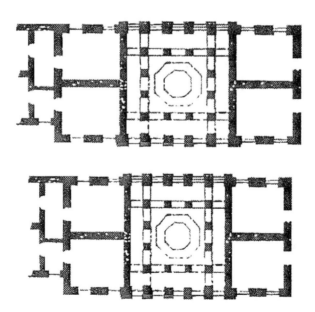

Figure 3. The plan of the first and the second storeys of the Admiralty tower. The architect I.K. Korobov (Museum of Leningrad History).

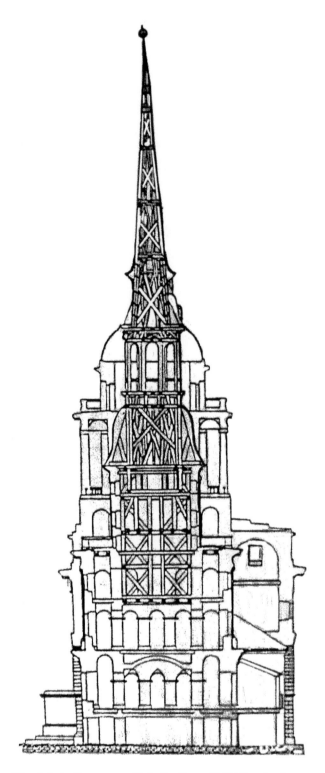

Figure 4. The cross-section of the Admiralty power with indication of I.K. Korobov's structures and the tower rebuilding by the architect A.D. Zakharov.

The building is located in the downtown city, in the River Neva delta, the part under restoration is located at the distance of about 500 m from the Neva embankment. The area is laid out, the absolute elevations of the ground surface range from +3 to +3.3 m BS. Within the investigation data under consideration (down to the depth of 35 m, which is lower than the depth of the active zone of impact of the existing structures),

Table 1. Soil properties.

No	Soil/layer thickness	Geological index	W	γ, kN/m^3	e	I_L	φ	c, kPa	E, MPa
1	Man-made soil, 3...5 m	tg$_{IV}$		16.5					
2a	Silty saturated sand with plant remnants, medium density, 0...1.4 m	ml$_{IV}$	–	19.4	0.75	–	25	2.0	18.0
2	Plastic clayey sand, 1.8...2.5 m	ml$_{IV}$	0.21	20.5	0.57	0.33	23	4.0	16.7
3	Clayey sand 7.5...7.9 m	ml$_{IV}$	0.27	19.7	0.73	1.29	19	11.0	10.9
4	Varved sandy clay, 14...17 m	lg$_{III}$	0.35	18.6	0.97	1.17	10	7.0	6.7
5a	Clayey sand with gravel and pebbles, 0...1.9 m	g$_{III}$	0.13	22.4	0.36	0.60	25	10.0	10.0
5	Sandy clay with gravel and pebbles with inclusion of sand	g$_{III}$	0.20	20.9	0.56	0.33	26	6.0	15.0

W – water content; γ – unit weight; e – void ratio, I_L – liquidity index, φ – friction angle, c – cohesion, E – deformation modulus.

the site is formed by modern marine and lacustrine deposits, Upper Quaternary lacustrine-glacial and glacial deposits covered with the anthropogenic layer. The total thickness of the anthropogenic deposits is up to 5 m, they consist of packed sands and construction debris.

At the Garden side within the upper 14–16 m (from the absolute elevations +1...−2 m to −10.7...−12.8 m BS) marine and lacustrine deposits consist of interlayering sands of different grain size (silty sands to coarse-grained sands with fine-grained sands domination) with inclusion of plant remnants, they protrude towards the Neva and the investigated building. They are underlain by clayey sands and sandy clays alternating in depth and extension. Their total thickness is about 7 m (with minor sand interlayers). The contractor found silty sands only in borehole 3 at the depth 3.4 m (the absolute elevation – 0.25 m), their thickness is 1.4 m. Clayey sands lie immediately under the anthropogenic layer nearer to the Neva. Within the upper 1.8...2.5 m (from the absolute elevations −1.65...−2 m to −3.8...4.2 m) the clayey sand is plastic. Saturated clayey sand with sandy clay interlayers lies below, down to the absolute elevation −11.7 m.

The roof of underlying Upper Quaternary lacustrine-glacial stratified varved sandy clays (mostly of liquid constituency) is at the depth of 14.7 m. Moraine sandy clays of hard plastic consistency, which are locally (borehole 3) covered with a 2-m-thick layer of hard plastic clayey sands), located at the depth 29...31.7 m (at the absolute elevations −26.3...−28.7 m BS).

Values of soil properties are given in Table 1.

The ground waters associated with saturated soils, sands and sandy interlayers in post-glacial clayey sands were encountered during investigations of different years at the depths −1...2 m, corresponding to the absolute elevations +2...+1 m BS.

The site is typified as follows:

1. There is a considerable (up to 5 m) layer of filled soil underlain by dense hard clayey sands (and silty sands) down to a depth of about 7 m.
2. There is a considerable (more than 20 m thick) bulk of saturated late glacial and post-glacial sandy clays and clayey sands, which are unstable due to dynamic impact (traffic, impact of metro lines) and highly compressible. Disturbance of natural structure due to dynamic impact leads to increasing deformability of soft soils and decreasing strength properties.
3. There is a high level of ground waters.

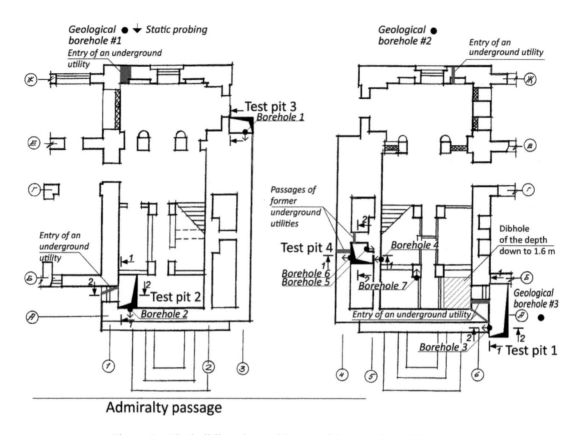

Figure 5. The building plan and layout of the test pits and boreholes.

4 EXAMINATION OF THE SOIL AND FOUNDATIONS

The present examination of the soil and foundations of the Admiralty main tower included the following works:

- digging, inspection and documentation of 4 test pits;
- drilling 7 boreholes through the foundations;
- inspection of wooden historical piles;
- sampling and studying soil specimens;
- dynamic sounding of soils;
- microbiological analysis of samples of pile timber.

Test pits were made inside and outside the tower building. The test pit inspection is described in Section 3.

Boreholes were drilled at the angle 10–15° from the vertical through the foundations body and inside the building. The results of drilling are given in the Section 3. Soil samples were taken from boreholes below the foundation base, they were studied in a geological laboratory.

A sample of a wooden pile was taken from borehole # 2, it was used for microbiological analysis in the laboratory of the Governmental company "Spetsprojectrestavratsiya". Dynamic sounding of soils was performed in the test pits (except test pit 4) in order to evaluate soil state under the foundations. A geodetic service carried out geodetic survey of absolute elevations of surfaces and foundations using a leveling device.

The layout of the test pits and boreholes made during the investigation is shown in Fig. 5.

4.1 Inspection of the test pits

Test pit 1 (Fig. 6) was made outside the building at the façade side along the side façade wall of the tower. The test pit was dug along the whole length of the side façade down to the depth −2 m from the ground surface. The façade wall of the Admiralty building is made of brick, it is plastered, there is no basement, and the

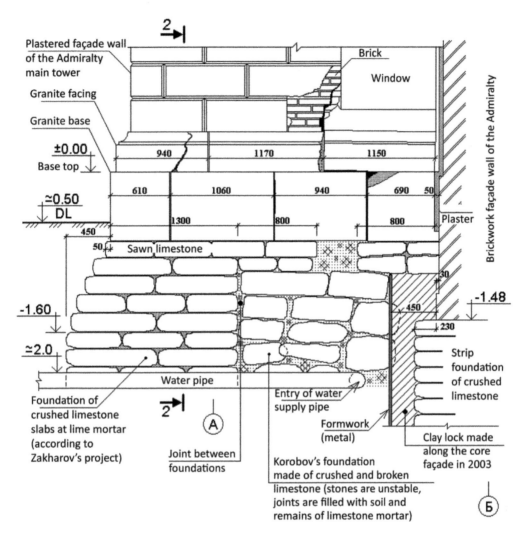

Figure 6. Test pit 1, cross-section.

brickwork is embedded in soil to a depth of about 1.0 m. The bottom of the embedded brickwork is covered with clay lock; above the plaster the brickwork is open and wet.

There is a strip foundation under the side wall of the tower. Under the façade wall it consists of two parts constructed in different years: the initial old foundation of Korobov's building located closer to the core wall, and Zakharov's foundation added from the front façade of the tower.

The foundation joint is located at the distance of 1.3 m from the façade corner of the base. Korobov's foundation within the test pit is made of different-sized crushed and broken limestone with remains of lime mortar and soil in brickwork joints. Zakharov's foundation consists of crushed limestone of 17–18 cm high at lime mortar.

Subsoil under Zakharov's foundation is composed of medium-grained sand (apparently, the artificial upfilling) of medium density identified during the probing. There is a strip foundation under the Admiralty core wall, it is made of crushed limestone with the edge protrusion of 23 cm. The edge is located at the depth 1.48 m from zero elevation.

The test pit identified a clay shallow excavation with vertical metal formwork made in 2002 along the façade wall of the building; it covers the upper part of the foundation and the lower part of the embedded brickwork of the wall.

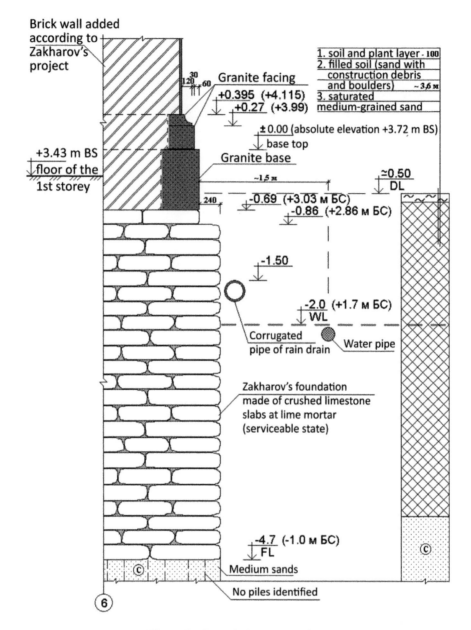

Figure 7. Test pit 1, cross-section.

Test pit 2 (Fig. 8) was excavated near inner wall of the left side of the Tower. There are strip foundations under the tower of Korobov's design. Height of the foundation strips is about 16–22 cm. During the borehole execution wooden piles were found, and pile material was taken for testing and investigation. The foundation rests on the layer of saturated fine sands and silty soils.

Test pit 3 (Fig. 9) was made outside the tower from the backyard side at the corner of the walls of the backyard and side façades. The test pit was dug along the whole length of the backyard wall up to the arch passage and along the sidewall to the doorsteps. The test pit opened the upper part of the foundation of the backyard wall and the foundation structure under the side wall, the foundation base was reached by prospecting drilling. Both brick walls are plastered. The backyard wall belongs to the initial Korobov's building. The side wall was added later according to Zakharov's project. The backyard wall has a basement consisting of 5 rows of 18-m-high sawn limestone slabs. The brickwork is wet; there are some blooms at its surface.

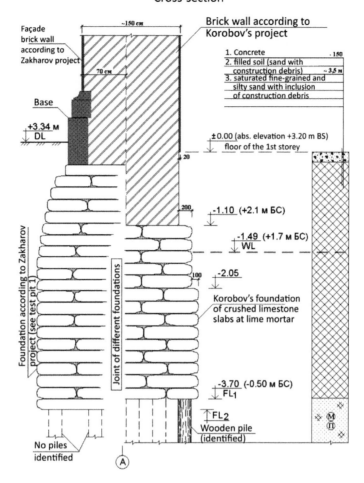

Figure 8. Test pit 2, cross-section.

Korobov's wall has a strip foundation consisting of crushed limestone slab at lime mortar. The foundation has rectangular cross-section with the edge protrusion of 15 cm. The foundation edge is located at the depth 0.88 m from the ground surface (the absolute elevation +2.05 m BS). The foundation base was found at the depth 3.78 m from the ground surface (the absolute elevation −0.85 m BS). The foundation height is 2.9 m.

The prospecting drilling has not encountered any wooden piles under the foundation. There is no water isolation along the foundation edge. Subsoil under the foundation consists of silty sand of medium density (identified by sounding). The added side wall does not have a rubble stone strip foundation. The foundation structure rests on an underground 2.5 bricks high arch (70 cm), at the test pit location it is embedded in Korobov's foundation made of rubble stone, the toe base is at the depth of about 1.5 m from the ground surface. Apparently, the other arch toe rests upon the rubble stone foundation of the newer Zakharov's building. The state of rubble stone masonry of the Korobov's foundation in the area opened by the test pit during the examination was found serviceable, without visible defects and damages, characterized by superficial washing out of the mortar. The brick arch foundation structure of the Zakharov's building is also serviceable, without cracks and joint opening.

Dynamic probing of soils was made in 4 locations, in the areas of open test pits. The sounding results showed that filled sands of different grain size, sandy clays and silty sands of natural sedimentation, alternating in the tower foundation subsoil, are dense and they have good strength properties. Filled soils above the foundations bases are less and unevenly compacted.

Despite its heterogeneous composition (of filled and natural origin) the tower foundation subsoil can be regarded as rather reliable with quite high soil density.

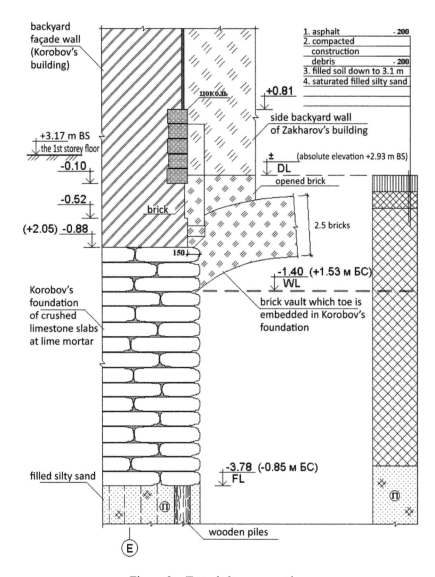

Figure 9. Test pit 3, cross-section.

5 EVALUATION OF THE STATE OF THE FOUNDATION VIA VIDEO RECORDING OF BOREHOLE CAVITY

In order to evaluate the state of the rubble stone foundation, the foundation body was surveyed by a mini TV camera within the drilled boreholes. Applying a mini TV camera allowed inspecting inside the foundation body identifying unfilled joints of rubblework, existing cavities and cracks.

The drilling results showed that the rubblework quality varies in different parts of the building. The video recording of the internal cavity of boreholes allowed visual identification of unfilled joints of rubblework, existing cavities (5–100 mm large), dips, failing joints between separate rubble stones in the upper part of the foundation. Video recording was performed down to the ground water level.

Fig. 5 shows the layout of the boreholes, Figs. 10–13 show the appearance and cross-sections of two boreholes.

Figure 10. Borehole 1.

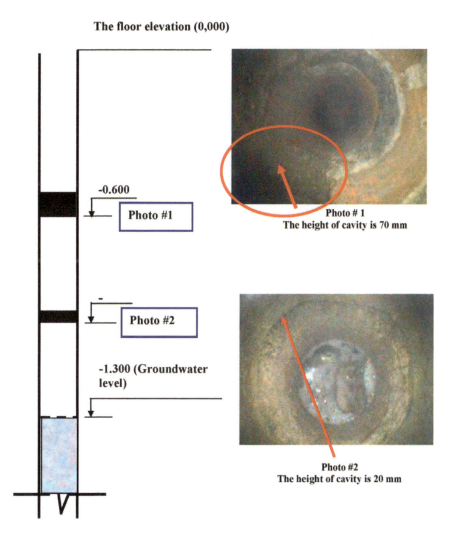

Figure 11. The scheme of borehole #1 and the fragments of video recording of cavities in the foundation body.

Figure 12. Borehole 2

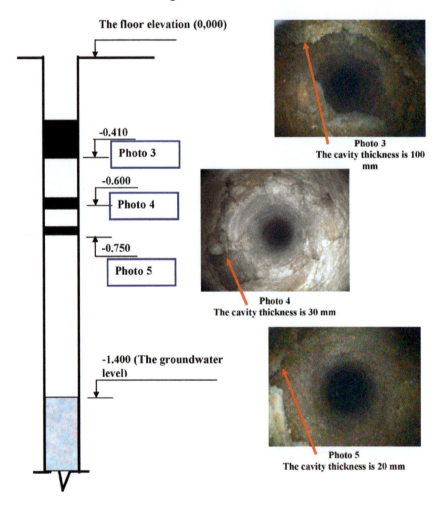

Figure 13. The scheme of borehole 2 and the fragments of video recording of cavities in the foundation body.

In borehole 1 the video recording was made down to the depth of 1.3 m from the soil surface. The video recording results identified two locations with cavities (Fig. 11):

– there is a large cavity of 70 mm high at the depth 0.6 m;
– there is a cavity of 20 mm high at the depth 1.1 m.

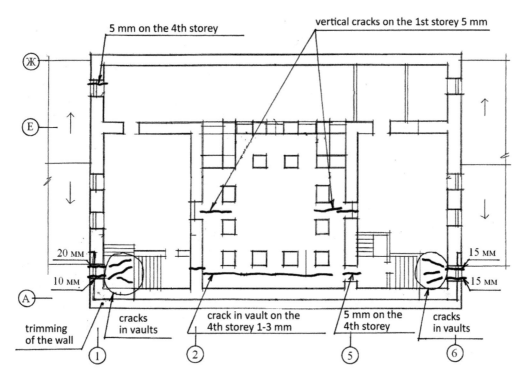

Figure 14. Location of cracks in plan. (The plan indicates the cracks located at different high level).

According to classification of the technical conditions in compliance with Construction Code 13-102-2003 and the results of video recording, rubblework is in serviceable state. The percentage of cavities in the borehole is 6.9%. At the moment of examination ground waters were encountered at the depth 1.30 m from zero elevation.

In the borehole 2 the video recording was made down to the elevation 1.40 m from the floor level. The video recording results identified three significant cavities as well as cracks (Fig. 13);

– there is a large cavity of 100 mm high at the depth 0.410 m;
– there is a cavity of 30 mm high at the depth 0.600 m;
– there is a cavity of 20 mm high at the depth 0.750 mm.

According to classification of the technical conditions in compliance with Construction Code 13-102-2003 and the results of video recording, rubblework is in partly serviceable state. According to the calculations the percentage of cavities in the borehole is 10.7%. At the moment of examination ground waters were encountered at the depth 1.40 m from the first storey floor elevation.

6 EXAMINATION OF THE TECHNICAL CONDITION OF THE ADMIRALTY SUPERSTRUCTURES

The authors examined the current technical condition of the bearing walls of the monument. The walls are made of brick with lime mortar. Up to this date the Admiralty tower walls were examined in detail by two companies in 2001 and in 2003, which compiled the schemes of cracks. In 2003 the values of additional crack opening were measured during three months. The measurement results showed that in the given period cracks opened and closed. The maximum width of additional crack opening was 0.18 mm.

During the examination analysis of the crack locations was conducted, with account of their influence on the condition of the whole building. The main cracks in the building plan are shown in Fig. 14. Fig. 15 presents the location of cracks in the side façade wall. Fig. 16 shows the location of cracks in the internal walls.

The location of cracks shows that the main façade wall along the axis "A" is separated by cracks from the other parts of the building. This defect was caused by increasing load applied to the wall during the Admiralty

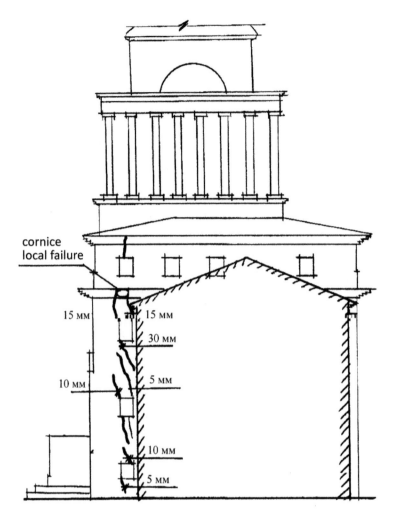

Figure 15. The scheme of cracks location in the wall along axis 6.

reconstruction in 1816. The identified nature of cracking is considered inside in the calculation scheme, which takes into account soil-foundation-superstructure interaction for the tower (see Fig. 21, 22).

7 NUMERICAL CALCULATION OF STABILITY AND DEFORMABILITY OF THE ADMIRALTY BUILDING

In order to calculate stability and deformability of the Admiralty building the authors made measurements of structures. Cross-sections and plans were elaborated to conduct numerical analysis of strain-stress behavior of the soils. Fig. 17 gives one of the examples of measurements. The green colour shows structural elements constructed in 1732 according to the project of the architect I.K. Korobov, the red colour shows structural elements constructed in 1816 according to the project of the architect A.D. Zakharov (Fig. 17).

In order to evaluate deformations of the building and subsoil the authors performed 3D Finite Element calculations using software *FEM-models 2.0* (Shashkin K., 2006). This software was developed by Georeconstruction Engineering Company.

Volumetric elastic elements were used for simulation of the building structures consisting of massive brick walls. This modeling allows identifying the heavily loaded parts of the masonry based on the elastic stage of the material behavior.

For modeling soil behavior authors used non-linear elasto-viscoplastic model with independent description of hardening during volumetric deformations and shear. This model provides obtaining good approximation of soft saturated clayey soils behavior. The model parameters were based on the data of geotechnical

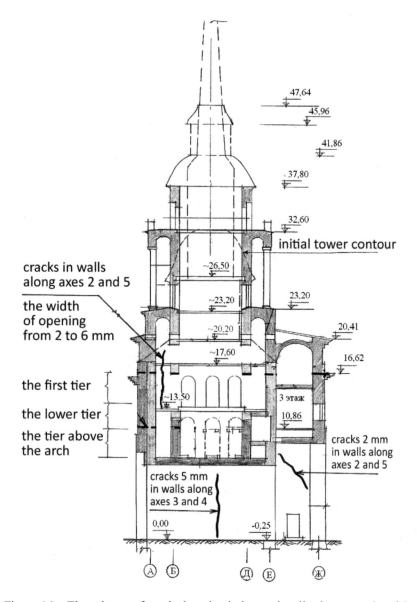

Figure 16. The scheme of cracks location in internal walls along axes 2 and 5.

investigations, including results of laboratory testing and CPT probing. The calculation scheme is shown in Fig. 18.

During the modeling of the monument's behavior authors tried to reproduce real structural scheme of the building. Also, during calculations history of the building construction was taken into account. On the first stage of analyses deformations of Korobov's Admiralty building (1732) were calculated. On the second stage of analyses deformations of Zakharov's Admiralty building (1816) were calculated, taking into account additional structures and loads. In Fig. 19 and 20 locations of masonry belonging to different stages are shown in different colours in the calculation scheme of the building.

Fig. 21 shows the summary contours of settlements after reconstruction of the Admiralty. According to the calculation results the total settlement is about 20 cm. It must be noted that the calculations do not take into account long-term soil creeping leading to the development of settlements at the constant speed of several mm per year. The applied elasto-viscoplastic model allows considering such deformations linked to relaxation of stresses in the subsoil and slow creeping of clayey soils.

The performed analyses of the building settlements development allowed identifying the causes of the emerged defects in the building structures. The main and the most pronounced defect is the development of

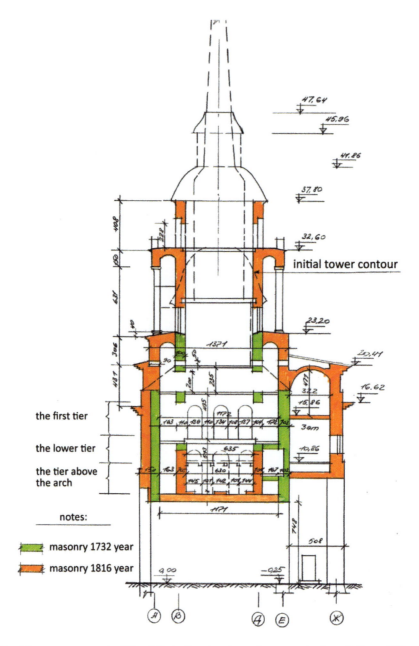

Figure 17. The measurement of the Admiralty structures for numerical modeling of the structure.

cracks in the wall along axes 1 and 6 between the axes A and B (see Fig. 14, 15). According to the calculation the central high-rise part gets the larger settlement than less loaded wings.

The wall along the axis A is linked with the walls along axes 2, 3, 4 and 5 (see Figs. 14–16). It has significant rigidity, its settlement corresponds to the settlements of the central part. At the same time the less loaded wall along the axis B develops a smaller settlement. As a result, there is a differential settlement in the wall along axes 1 and 6 between the axes A and B, which is about 3 mm according to the calculation. There is a small difference of settlements in the elastic scheme. However, difference of settlements leads to the development of considerable shear stresses along axes 1 and 6 (Fig. 22). This corresponds to the shear of wall along the axis A in relation to the wall along the axis B. Shear stresses reach the value of 235 kPa. This leads to emergence of inclined cracks in the walls along axes 1 and 6. Opening of these cracks upwards can be explained by some tilt of the tower. The calculated tilt towards the street façade is a result of the asymmetric location of the tower against the building (about 16 mm according to the top of tower).

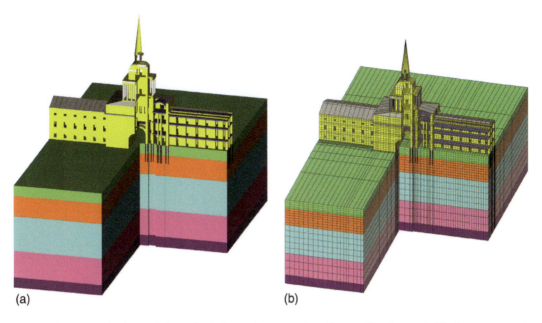

Figure 18. (a) The general view of the calculation scheme with soil stratification and (b) the cross-section of the Admiralty structures.

Figure 19. Cross-sections of the building. The masonry of 1732 (architect Korobov) is indicated in green, the masonry of 1816 (architect Zakharov) is indicated in orange.

The calculation also identified a zone of the possible cracks development in the walls of the wings. The reason is differential settlement of the central part and the wings. Calculation of structures in elastic domain obtained a significant area of the principal tensile stresses (Fig. 22), showing a potential possibility of cracks development. However, it does not mean that the cracks should occur all over the area, as during cracks

Figure 20. Cross-section of the building. The masonry of 1732 (the architect Korobov) is indicated in green, the masonry of 1816 (the architect Zakharov) is indicated in orange.

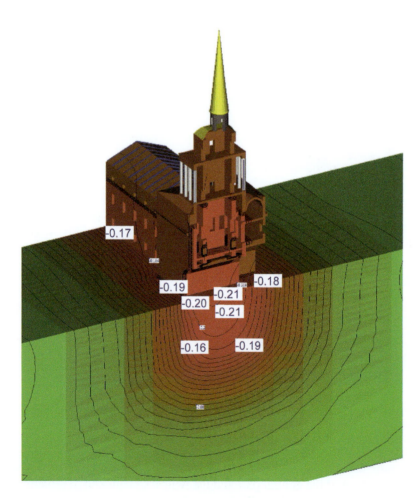

Figure 21. The calculated contours of settlements (m) of the Zakharov's Admiralty structures (the cross-section along the symmetry axis).

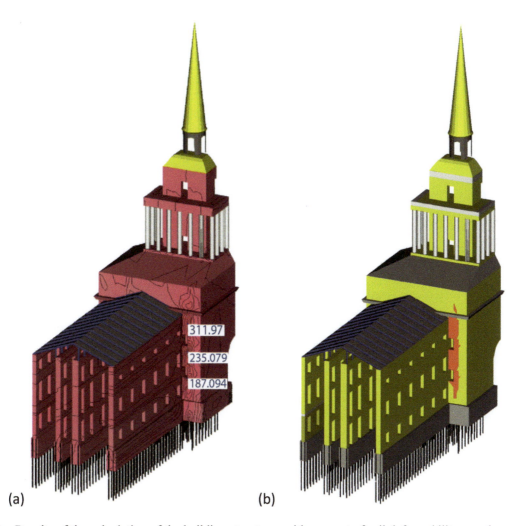

Figure 22. Results of the calculation of the building structures with account of soil deformability: a – the contours of shear stresses in the masonry (kPa) caused by differential settlements of the walls of the wings and tower; b – the areas of possible development of the inclined cracks associated with shear deformations of the transversal wall.

emerging there is a big change of the strain-stress state and tensile stresses disappear. During the examination of the structure the authors have not registered significant development of cracks in the wings. Probably, the cracks, which were caused by the settlements, emerged in the initial period of the building existence and they were plastered and did not get further opening.

Therefore, the differential settlements of the building connected with the large load in the central part of the building are the reason of the main defects of the Admiralty building (cracks in the walls along axes 1 and 6 between the axes A and B). Settlement development can be currently ongoing, due to soil creeping. Currently these deformations are not so well studied, and they have not been taken into account in the performed analyses. It is advisable to organize monitoring of the development of the building settlements in time and opening of the existing cracks. This will give us possibility to make a conclusion about the danger of development of further defects in the building structures.

8 CONCLUSIONS

The deformations of the Admiralty are caused by historical rebuilding of the structures, use of different material incompatible with the monument, changes of the loads applied to separate foundations, decay of wooden piles and log rafts constructed in compliance with the Regulations of the time. The final factor is associated with the lowering of ground water level, canalization and drainage of the area. We have established that the Metro also influences the deformation development.

The reader would be pleased to know if there is claiming for any remedial measures or if, once the causes of damage have been identified, we can rely on the actual state by just limiting remedial measures to plastering the fissures. In this way, this case history could be an example that show how a detailed survey of the structure, foundation and soil can be beneficial in avoiding to rush towards unnecessary intrusive interventions, but rather suggesting the implementation of a monitoring programme to investigate the evolution of settlements (if any) or developments of fissures and cracks.

REFERENCES

Burghignoli, A., Jamiolkowski, M., and Viggiani, C. (2007). Geotechnics for the preservation of historic cities and monuments: components of a multidisciplinary approach. General report. *Proceedings of XIV European conference on Soil Mechanics and geotechnical Engineering "Geotechnical Engineering in Urban Environments"*, Madrid, Volume 1.

Shashkin, C. (2006). Basic regularities of soil-structure interaction. *Proceedings of XIII Danube European Conference on Geotechnical Engineering*. Ljubljana. Vol. 1, pp. 179–190.

Ulitsky, V. (2003). Geotechnical challenges in reconstruction of historical cities (as may be illustrated by St. Petersburg). *Proceedings of the geotechnical conference "Reconstruction of historical cities and geotechnical engineering"*. St. Petersburg, 16–18 September 2003. Vol. 1, pp. 13–28.

Geotechnics and Heritage: Historic Towers – Lancellotta, Flora & Viggiani
© 2018 Taylor & Francis Group, London, ISBN 978-1-138-03272-9

Preservation of the main tower of Bayon temple, Angkor, Cambodia

Y. Iwasaki
Geo Research Institute, Osaka, Japan

M. Fukuda
Taisei Geotech Co., Kurume-city, Fukuoka, Japan

ABSTRACT: This chapter describes techniques of preservation of the main tower of the Bayon temple of Angkor Thom in Cambodia. The Japanese Government Team for Safeguarding Angkor was established in 1994 for the purpose of conservation work in the Angkor Archaeological Park. In this chapter, geotechnical and foundation studies of the main tower of the Bayon site are summarised with the emphasis on geotechnical engineering. The main tower of 31 m in height was constructed on a man-made sandy mound, 14 m in thickness, by direct foundation. Beneath the main tower, there is a vertical shaft in the foundation element that was backfilled with loose sand. The stability of the tower comes from the unsaturated, compacted soil mound. The character-defining elements are discussed in terms of the authenticity and stability of the foundation of the Bayon monuments.

1 INTRODUCTION

After the chaos of Cambodia's Pol Pot regime, in 1993 the Japanese Government hosted an intergovernmental meeting with UNESCO in Tokyo on Safeguarding Angkor. Historically, until the 1970s, the École Françoise d'Éxtrême-Orient (EFEO), France, had been working for conservation of Angkor. The 1993 meeting was to create an international framework through UNESCO for conservation of Angkor monuments. The Japanese Government Team for Safeguarding Angkor (JSA) was established in 1994 and started restoration and scientific research activities. In 2005, JSA was reorganised as a joint body of the APSARA national authority (Authority for the Protection and Management of Angkor and Region of Siem Reap) and the organisation was renamed as JAPAN-APSARA Safeguarding Angkor (JASA). The author was invited to join JSA in 1994 and has been engaged in geotechnical engineering for conservation work for more than 20 years.

2 ANGKOR AND BAYON, ANGKOR THOM

Figure 1 shows the location of Angkor to the north-east of the huge lake of "Tonlé Sap", which is the biggest lake in Southeast Asia, with very shallow water depths of approximately 3 m during the dry season, deepening to 7–10 m in the rainy season due to water backflow from the Mekong River at the river junction point at Phnom Penh. Angkor is primarily represented by Angkor Wat and Angkor Thom. However, many heritage temple structures are widely distributed between the northern shore of the Tonlé Sap lake and the top of the Phnom Kulen mountain range in an area 50 km in width, as shown in Figure 2.

Kulen Mountain is 700 m in height and has a unique bowl shape that is a natural water tank and provides water to the Siem Reap River year around. Figure 3 shows a plan view of Angkor Wat and Angkor Thom, which are located on flat ground approximately 25 m above sea level at the midpoint between Tonlé Sap lake and Mt. Kulen.

They are surrounded by moats that are one of the characteristic landscape elements of temples in Angkor. Bayon temple is located at the centre of Angkor Thom and was constructed in the late 12th or early 13th century by king Jayavarman VII (1185–1220).

In Khmer, Ba Yon means 'beautiful tower', and in Sanskrit it means 'thrones for gods'. The Bayon temple was constructed in a Buddhist style. Many four-faced towers surround the main tower, as shown in Figure 4.

Figure 1. General map of Cambodia.

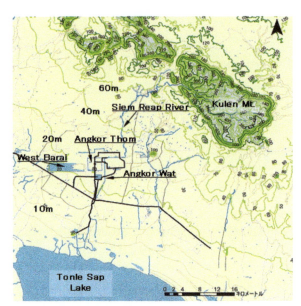

Figure 2. Angkor and its vicinity.

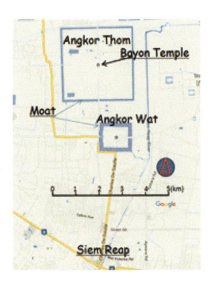

Figure 3. Plan view of Angkor Wat and Angkor Thom.

Figure 4. Bayon temple, Angkor Wat.

3 BAYON

A general plan and section view of Bayon temple are shown in Figure 5. The temple complex, which covers 144 m by 228 m, consists of an outer gallery, inner gallery, northern library, southern library, four-faced towers, and the main tower.

Figure 6 shows a general view of Bayon temple from the north-west direction. The main tower stands at the centre of the temple and is considered as the symbol of Angkor Thom. The main tower was surrounded by four-faced towers, as shown in Figure 7.

The corbel arch seen at the lower right corner of the figure was used to create a vault structure and is considered characteristic of stone masonry in Angkor. Bas-reliefs of war battles, as well as daily life and stone working, are sculpted on the walls of the outer and inner galleries of the temple, as shown in Figure 8.

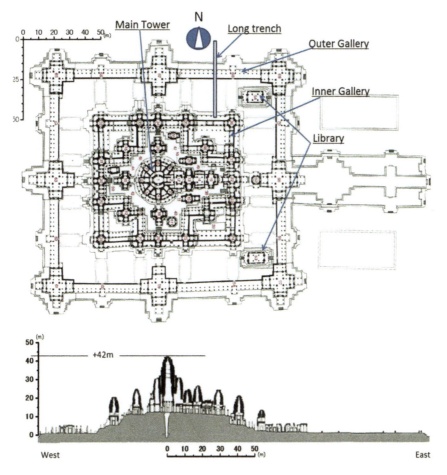

Figure 5. Section and plan of Bayon temple.

Figure 6. Bayon temple from the north-west direction.

Figure 7. Four-faced towers.

Figure 8. Bas-reliefs of daily life (left) and stone working (right).

4 MAIN TOWER OF BAYON

The foundation of the Bayon complex is made of three stepped terraces. The outer gallery, the northern library, and the southern library were constructed on the first terrace, 2.5 m in height from the ground. The inner gallery and the four-faced towers are on the second terrace of 6 m height. The main tower and further four-faced towers stand on the third terrace of 12.4 m height. Figure 9 shows a general view of the main tower from the north-west side. The main tower consists of the central tower and eight sub-towers that stand around the central tower. The north–south section and plan are shown in Figure 10. The main tower and sub-towers are constructed with masonry stone walls.

The main tower is considered to have been constructed in three stages, with the lower part of the central tower as the first stage, the sub-towers and mid-part of the main tower as the next stage, and the final stage being the top part of the central tower. A front chamber with a vaulted corbel structure is connected with the lower part of each sub-tower. A face was engraved on the outside of the top of each sub-tower with a lotus flower design at the top.

5 PRESENT STATE OF THE STRUCTURE OF THE MAIN TOWER OF BAYON

Figure 11 shows the top of the central tower which lacks the original symmetry that can still be seen in the four-faced tower shown in Figure 12. This lack is considered to have been caused by the collapse of stones on the outer surface of the tower.

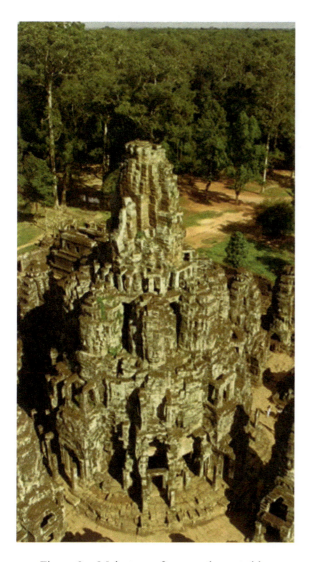

Figure 9. Main tower from north-west side.

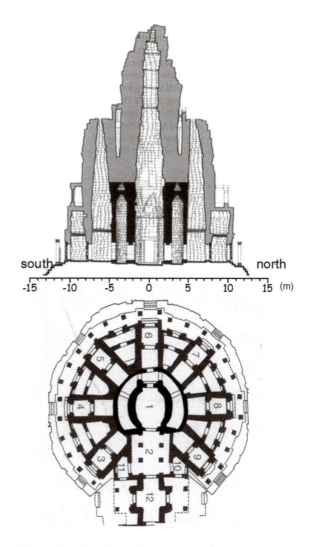

Figure 10. Section and plan views of the main tower.

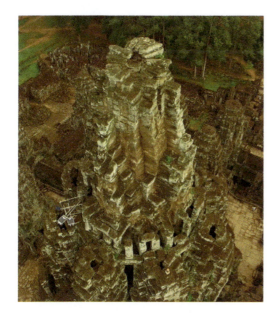

Figure 11. Top of the central tower.

Figure 12. Four-faced tower in Bayon.

Figure 13.   Plant at Ta Prohm temple.

Figure 14.   Plant root at the main tower.

Figure 15.   Crack on surface of column stone of the central tower in the mid-level gallery on west side.

Figure 13 shows an example of a tree that is growing over the masonry structure at Ta Prohm temple, where some plants have been allowed to grow as freely as before. Figure 14 shows a root growing in a gap between stones and it is anticipated that this will widen the openings between the stones.

There is a narrow space between the central tower and the sub-towers at the mid-level of the main tower, which is the mid-level gallery. Here, we found cracks on the surface of the stones of the tower masonry at several points, one of which is shown in Figure 15.

In the same mid-level gallery, two other cracks were also identified on the outer stones of the masonry columns of the central tower, as shown in Figure 16. If such cracks grow and reach the edge of the stone, the cracked stone becomes unstable. The method used to stack cut stones for the main tower in Bayon is a simple piling of independent columns without interlocking with adjacent columns, as illustrated in Figure 17. It is

Figure 16. Cracks on the surface of column base.

Figure 17. Simple stack of masonry stones.

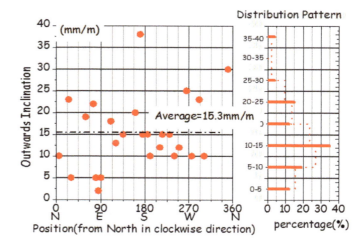

Figure 18. Inclination of side column of front chamber.

Figure 19. Distribution of inclination angles.

easy for a simple piling of stones to fail and this might have resulted in collapse when some stones began to break down.

The side columns of the front chambers of the sub-towers were found to be inclined outwards, as shown in Figure 18. The average outwards inclination was 15.3 mm/m, as illustrated in Figure 18. Inclined columns were found in many orientations of the sub-towers, as shown in Figures 19 and 20.

The front chambers have common walls with the sub-towers. Vertically continuous horizontal joints are found in most of the stone masonry walls. The gap width increases with height from the base foundation of

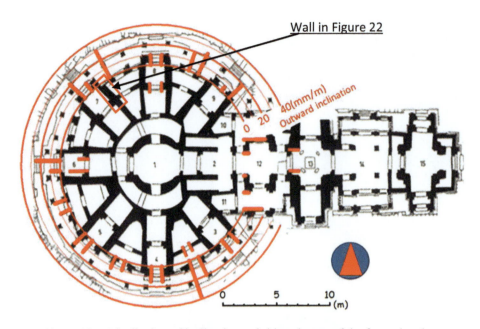

Figure 20. Distribution of inclinations of side columns of the front chamber.

Figure 21. Joint width increasing with height.

the wall, as illustrated in Figures 21 and 22. The average width of the gap per metre of height is about 10 mm. The largest width of gap is found in the north side wall of the sub-tower at the NW position (see Figure 20). Figure 22 shows the distribution of the gap widths on the wall surface. The rate of increase of the width of the joint gap is constant at a value of 15 mm/m from the base to a height of 3.5 m, above which it increases to 120 mm/m.

The structural deformation of the outwards inclination of the side columns of the front chambers, as well as the increase in the gap opening up between stones with increasing height, suggest tilting of either the stone foundations and/or the upper stone structures.

There are crossing corridors in the east–west and north–south directions in the main tower. The height levels of the surface of the base frame foundation stones at the boundaries between chambers were measured and are shown in Figure 23. The load of the upper stones is transferred by the side stone frame columns to the

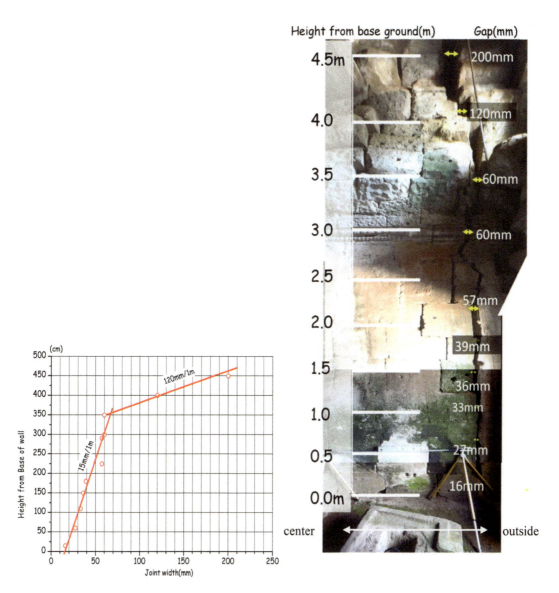

Figure 22. Increase of the joint width with height from the base (NW sub-tower).

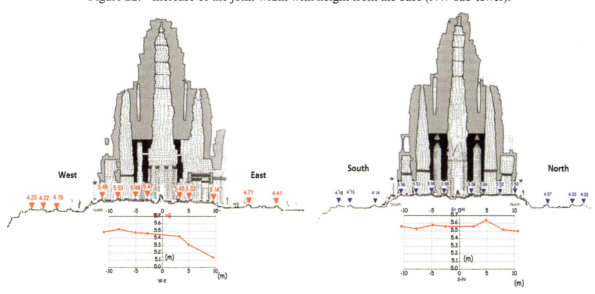

Figure 23. Height levels of the surface of the basement frame stones of the entrances for chambers.

base stone of the entrance frame supporting them. Thus the level of the surface of the base stones and the pattern of the levels is expected to show the basic characteristics of settlement of the foundation of the main tower. As previously indicated, the average inclination of the side columns of the front chambers is 15 mm per 1 m. Of the four orientations, the outwards tilting is found 50 mm, 20 mm, 4 mm per 1 m in the eastward, the westward, and the northward directions respectively. The tilting in the south direction is not the outwards but inwards with an inclination of 10 mm per 1 m. The inclination of the base frame foundation is found little tendency to be related with the outwards inclination of side columns.

In 1933, Mr. Georges Trouvé, then the conservator of EFEO, excavated the centre of the base ground, expecting buried treasures. He found a Buddha statue that had been broken into several pieces (see Figure 24(a)) and other architectural elements. The fragments were found between one and five metres below the ground surface. In fact, other excavations had previously been conducted and the broken statue of the Buddha had been discarded in the resulting hole. Trouvé continued to dig to 14 m below ground level where he stopped excavation because he encountered underground water. He also dug horizontal shafts at 12.5 m below ground level in east, south and west directions and found some stones that might have been a part of the inner structure of an ancient well. He speculated as to the existence of a well structure that had been constructed to provide a space for treasures and he suggested further excavation. However, due to the appearance of water at the bottom of the shaft, he halted the excavation, backfilled the shaft with soil, and the mystery remains. The level of the top of the shaft is ground level (GL) + 12.95 m; the bottom of the shaft is GL − 1.05 m, as shown in Figure 24 (b) and (c). At the time the shaft was excavated, it was September the rainy season and the underground water level was rather high; it may be lower in the dry season.

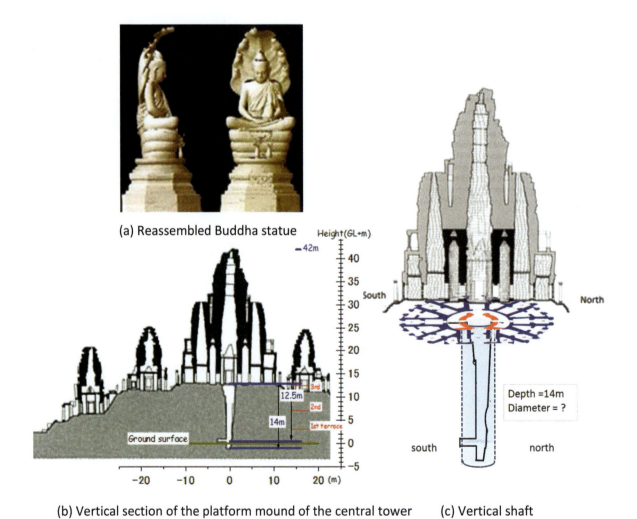

(a) Reassembled Buddha statue

(b) Vertical section of the platform mound of the central tower   (c) Vertical shaft

Figure 24. Vertical shaft under the main tower.

The present state of the structural aspects of the stone masonry 600 years after the construction of the temple in the late-12th/early-13th century can be summarised as follows:

1. Lack of symmetry at the top of the central tower that has been caused by the collapse of cut stones from the outer surface of the original construction.
2. Outwards tilting of the side columns of the entrance frames of the front chambers of the sub-towers.
3. Horizontal gaps in the stone masonry walls that continue vertically with increasing gap widths.
4. The existence of a vertical shaft under the main tower without lining that had been excavated either at the time of the construction or thereafter and was rediscovered in 1933.
5. The foundation of the main tower shows no apparent problems such as excessive settlement or excessive tilt of foundation other than the problems of tilting of the side columns of the front chambers and the horizontal masonry gaps (as in 2 and 3 above). However, these phenomena could be caused by tilting of the foundations.

6 GEOTECHNICAL CONDITION IN BAYON

This section describes general geotechnical conditions in Angkor, including Angkor Thom, before considering the stability problem of the main tower of the Bayon temple.

6.1 Geotechnical study of Angkor

In 1994, JSA commenced activities for a conservation project. The author visited the library of EFEO in Phnom Penh to learn about the past study of the Angkor heritage in the fields of geology and soil mechanics. Unfortunately, no study report on geology or geotechnical engineering was found in the available documents and reports. We started our geological study by creating survey boreholes of up to 100 m in depth to gain information about the ground condition in Angkor and reach base rock. The geological and geotechnical borings were performed at several points: in Bayon temple, the Royal plaza of Angkor Thom, Angkor Wat, and in Siem Reap city. The results are shown in Figure 25 with Standard Penetration Test (SPT) N-values.

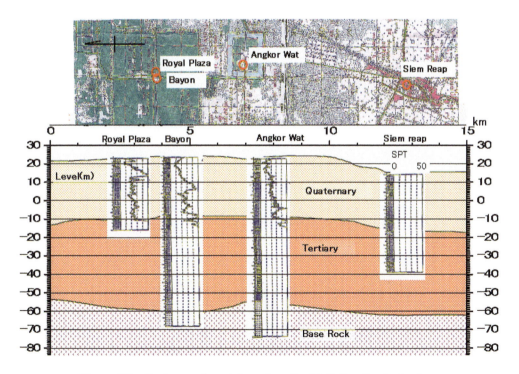

Figure 25. Boring sections taken along the N–S direction in Angkor.

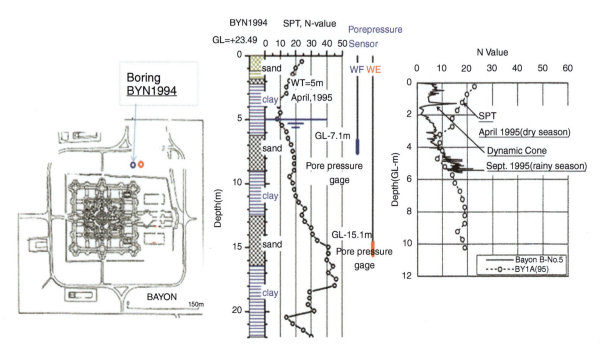

Figure 26. Boring results of BYN1994 at the north yard of Bayon temple.

The ground basically consists of three formations: quaternary rock at the top of 30–40 m thickness, then tertiary volcanic stone of 40–50 m thickness, followed by base rock. These geological formations have been found common to the Angkor plain by the later study of additional borings. The soils of the top quaternary period are estimated as a deposited formation that originated from the weathered sandstone formation of Kulen Mountain to the north of the Angkor plain.

6.2 *Geotechnical study of Bayon site*

The upper part of the long borehole at Bayon was made in the northern yard outside the Bayon temple and named as BYN1994, as shown in Figure 26 (a). The quaternary formation is about 30 m in thickness with SPT N-values of 5–50 from the alternation of sand and clay; the upper twenty metres is shown in Figure 26 (b). After confirming two sand layers in the quaternary formation, the installation of two pore water sensors in these sand layers was planned. An additional shallow boring was carried out and the sensors were installed in each of the two sand layers, one in each of the two boreholes (see WF and WE in Figure 26 (b)).

The underground water level was 5 m below ground level when the boring was performed in the dry season of April 1995, as shown in Figure 27. A dynamic cone test was performed at the same boring site in the rainy season in September 1995. The values equivalent to SPT, based upon the dynamic cone penetration test, are shown in Figure 26 (c). The resistance to ground penetration by SPT and dynamic cone testing show different trends in the top ground surface. The ground resistance at the shallower depth in the dry season has become much weaker in the rainy season. The difference is due to the change of underground water level, which was around GL − 5 m in the dry season and has risen to the level of the ground surface (GL − 0 m). The sand in Angkor is very sensitive to water conditions. The sampled soils of boring BYN1994 were tested to obtain natural water content and particle grain-size distributions, with the results shown in Figure 27. The ground consists of alternative layers of clayey sand, clay, and gravel soils. The most typical soils are generally clayey fine sand and stiff clay, with water content of about 10–20%.

Pressure sensors were installed at two different levels, GL − 7.1 m and GL − 15.1 m, to monitor the changes of the surface of the underground water at the site. The monitored data for the water pressures are shown in relation to hourly rain in Figure 28. Water pressures at the two different levels show different responses to the rain water. The water level at the shallow depth responds to rainfall rather quicker than that at the greater depth. Underground water in the Angkor area is being pumped out on a daily basis for use by local people.

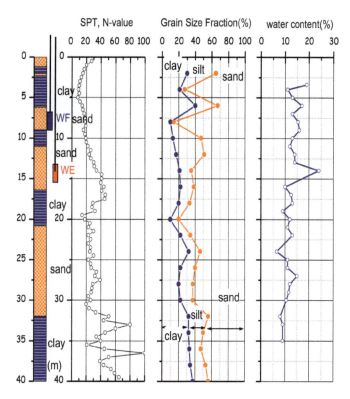

Figure 27. Soil properties at BYN1994.

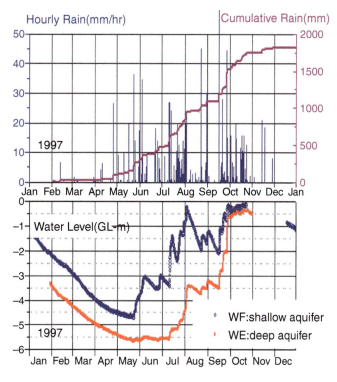

Figure 28. Underground water and rain at BYN1994.

The screen of the pumping pipe is commonly installed at a depth of around 20–30 m below ground level. The water level in the deeper sand layer is thought to be affected by the human activity of pumping water. The annual change in water level is found to be around 5 m, with the lowest level in May and the highest in November.

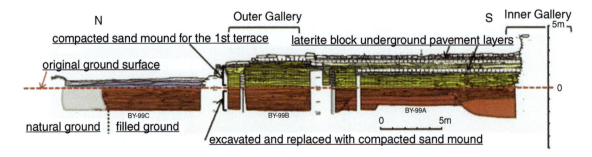

Figure 29. Vertical section of the long trench at north side of Bayon (Narita, 2000).

7 FOUNDATION SYSTEM OF BAYON TEMPLE

Bayon temple is a complex of stone masonry structures which were constructed on three stepped terraces. At several points, the terrace structure was studied by EFEO by means of archaeological trenches. The results showed a man-made ground of sandy soil interbedded with laterite blocks.

7.1 Trenched foundation

JSA made an extensive archaeological trench from the north side of the inner gallery in a northerly direction beyond the outer gallery. The location of the trench is marked in Figure 5 as 'Long trench'. The results from the trench of some 35 m in length are shown in Figure 29.

The archaeological trench has revealed the process of formation of the sand mound of the foundation. As the first step, the natural ground was excavated 2–3 metres in depth, then replaced with sandy soil and the fill was compacted. This trenched foundation was identified in the wider area as extending about 10 m outside the boundary of the masonry structure of the outer gallery of the Bayon temple. The top of the filled terrace is paved with a sandstone layer followed by laterite blocks beneath. Another layer of laterite blocks was found about 1.5 m below the surface. The filled sand was found to have been compacted layer by layer.

7.2 Three-stepped terrace of soil embankment

Based upon the results of the archaeological trench, the foundation of the Bayon temple is considered to have been constructed according to the following steps, shown in Figure 30. First, the original natural ground was excavated 2–3 m from the surface. Then, the site was backfilled with borrowed material of sandy soil with compaction. The first, second, and top third terraces were constructed with ground surfaces of +2.5 m, +6.0 m and +12.4 m, respectively. The surfaces of these terraces were paved with laterite blocks covered by sandstone blocks.

7.3 Foundation of the main tower of Bayon

An archaeological trench was excavated to study the foundation structure of the main tower in 2008. The masonry structure of the central tower is supported by four blocks of sandstone of semi-circular shape, as shown in Figure 31. The inside of the central tower is the main chamber of Bayon temple where the broken statue of Buddha was found. After removing surface stones in the northern floor of the main chamber, the north-western side was excavated to study the structure of the foundation. The foundation found at this north-western corner was a rather simple structure of sandstone of 20 cm in thickness followed by one or two layers of laterite block, as shown in Figure 32. This simple foundation was constructed upon a densely compacted sand fill mound.

As noted earlier, Trouvé reported that he had found some further stone structures at a horizontal distance of 1.20 or 1.75 m from the wall of the shaft of 1.2 m in diameter at a depth of 12.5 m from the ground of the chamber of the main shrine. The stone structure might be a part of a bearing wall to support the masonry structure of the main tower (Nakagawa et al., 2014). If the stone structure is a part of the foundation, it should continue upwards to the bottom of the foundation stones of the central tower.

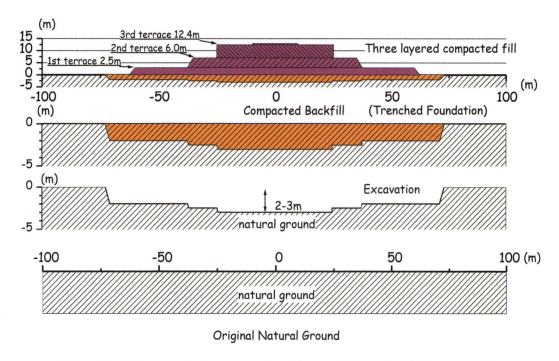

Figure 30. Three steps in the construction of terraced foundation for Bayon temple.

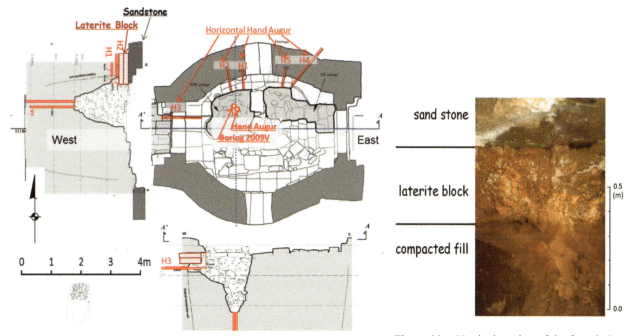

Figure 31. Foundation of the central tower at the main chamber.

Figure 32. Vertical section of the foundation at NW corner.

To confirm the materials under the laterite blocks, an additional survey was planned. Horizontal hand-auger sampling at depths below the laterite block was carried out at five different locations inside the excavated wall, as shown in Figure 31. However, nothing but the compacted sandy fill was found in all auger samplings (Shimoda et al., 2009).

Amazingly, the foundation is a shallow direct foundation system. Generally, the stone structure of Angkor is found to stand upon a shallow foundation of base stone and laterite block. The foundation system of the central tower of Bayon was identified as being the same as others in Angkor.

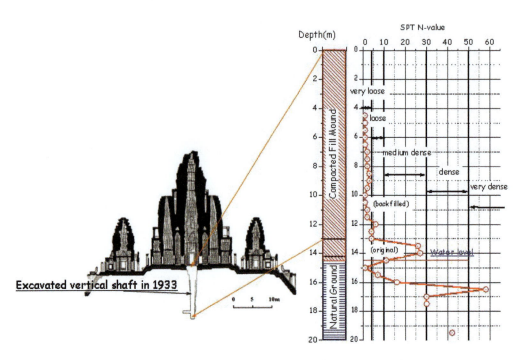

Figure 33. N-value results of SPT of boring BYC2009 at the vertical shaft at the central tower.

7.4 Vertical shaft in the main chamber of the central tower

After the original excavation of the vertical shaft, it had been backfilled. The diameter of the shaft is estimated to be as large as 2 m when space to dig the soil and carry it to the surface is considered. In 2008, a vertical hand-auger sampling of the north-western side of the ground of the main shrine was performed and the backfilled soil was found to be in a very loose state. To obtain more detailed geotechnical information, a boring (BYC2009) was carried out at the north-western side of the filled ground of the vertical shaft, as shown in Figure 33, in which the N-value results of the SPT are also shown. It became clear from the SPT that the backfilled soil is very loose, with N-values below 4 from the surface down to GL − 12 m. Natural ground was identified at GL − 14 m (Iwasaki et al., 2014).

The vertical shaft at the centre of the main tower of Bayon temple is considered to have been excavated as a result of religious conflicts between Buddhism and Hinduism. The Bayon temple was constructed around the late 12th century by king Jayavarman VII and based upon Buddhism. However, his successor, king Jayavarman VIII, is known to have believed in Hinduism and had broken statues related to Buddhism. It is not known when the vertical shaft was excavated; however, it might not have been long after the death of king Jayavarman VII.

It is rather amazing that the very heavy stone central tower structure of more than 30 m in height has been stable for several centuries. Even without the vertical shaft at its centre, present levels of geotechnical engineering could not provide a solution of direct foundation for such a heavy stone structure directly on man-made fill of 14 m in height without soil improvement or the use of geotextiles.

7.5 Borings at the foundation platform

After the BYC2009 boring at the vertical shaft in the main chamber, JASA started geotechnical study from the third terrace surface with several borings in vertical and horizontal, as well as inclined directions, as shown in Figures 34, 35, 36, and 37. BYV2010 was the only vertical boring with SPT N-values. Other borings were either horizontal or inclined borings to study the geotechnical structure of the foundation mound.

The basic structure of the mound was obtained from BYV2010, which shows the weathered laterite blocks of about 6 m in thickness beneath the sandstone and laterite pavement and stiff compacted sand layers down to the bottom of the mound. From the same starting point as BYV2010, two inclined borings – BYH2010-30 and BYH2010-45 – were performed at 30 and 45 degrees, respectively, in a northerly direction. As shown in Figures 35 and 37, weathered laterite (red color) was identified from the top of the mound beneath the

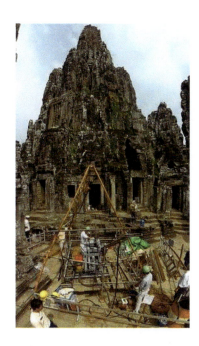

Figure 34. Boring BYV2010.

Figure 35. Borings at the mound of the main tower (plan).

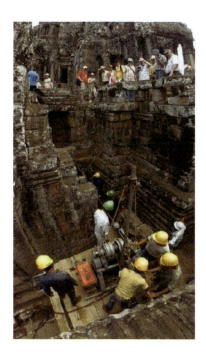

Figure 36. Boring BYHN2012.

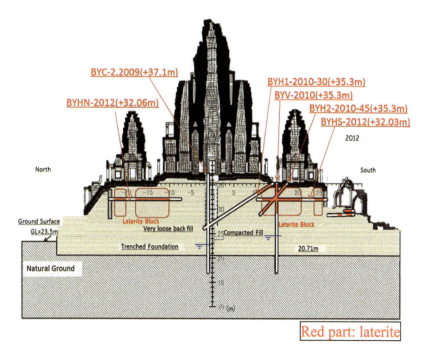

Figure 37. Borings at the mound of the main tower (section).

pavement in all three borings in 2010 as about 6 m in thickness and about 5 m in horizontal width. The image of the existence of a large laterite body emerged at the outer periphery of the foundation mound of the central tower. In 2012, two additional horizontal borings, of BYHS2012 from the south and BYHN2012 from the north, were carried out and also gave the results expected for the existence of this laterite body.

After the construction of the second terrace with a surface at GL + 6.0 m, the third terrace was expected to have been constructed by stepwise construction, with laterite blocks being laid at the outer periphery and sand material placed in the central zone inside the laterite blocks of about 10 m in width. Within each step, the sand layer had been compacted every 10–15 cm and this continued to the top of the third terrace of 6 m in height.

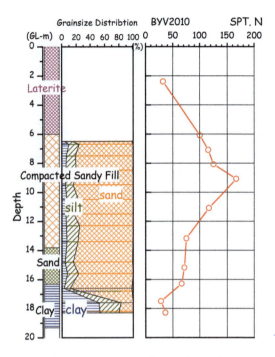

Figure 38. Boring log BYV2010.

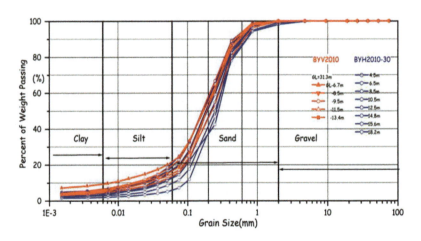

Figure 39. Sample sand at GL − 8.5 m, BYV2010.

Note too that the diameter of the central shaft, as a length of sample of very loose material between stiff sand, was estimated as 2.7 m by the inclined boring BYH2010-30.

7.6 *Geotechnical characteristics of the compacted sandy mound*

7.6.1 *Grain-size distribution and standard penetration test N-values*

Figures 38 and 39 show the N-value results of the standard penetration tests and the grain-size distribution of the obtained samples for the vertical boring of BYV2010.

The grain-size distributions of the sampled soils of compacted fill of the vertical boring of BYV2010 and the inclined horizontal boring of BYH2010-30 are shown in Figure 39. The grain-size distribution shows very little content of gravel-size grains in the fill. Common N-values for compacted sand fill without gravel lie within a range of 20–30. An N-value larger than 50 is very rare for sandy filled soil. The sampled soil looks like very stiff soil or soft sandstone. Very large N-values of more than 50 for sandy filled ground would involve special mechanisms such as treatment by mixing with lime.

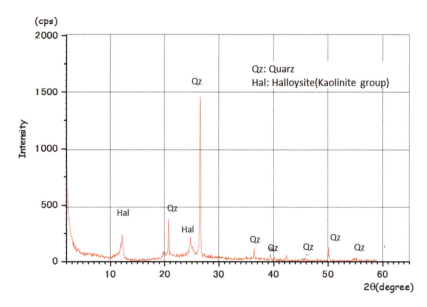

Figure 40. X-ray diffraction analysis for stiff sand.

Figure 41. Photomicrograph of the sand.

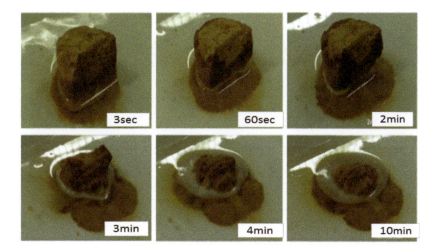

Figure 42. Response of the stiff sand to standing water.

7.6.2 *X-ray diffraction analysis*

X-ray diffraction analysis for a grained specimen of the stiff sand was performed and the result is shown in Figure 40. Figure 41 shows the original state of the sand sample. Each grain is recognisable but seems to stick to others. The X-ray diffraction shows two kinds of mineral components: quartz and halloysites. Halloysites belongs to the kaolin group of clay minerals; the sand is identified as kaolin sand, which shows special characteristics of great strength under dry conditions.

Kaolin clay provides very strong cohesion caused by large negative suction pressure under unsaturated conditions. However, when kaolin sand is submerged in water, the negative suction will be lost and the strong cohesion begins to disappear (Iwasaki *et al.*, 2014).

7.6.3 *Response to submerged water conditions*

Figure 42 shows the response of the stiff sand of the sampled soil to water.

The successive photos show how the soil collapses when standing in water: within 10 minutes the stiff sand had collapsed and the block had disappeared in the water.

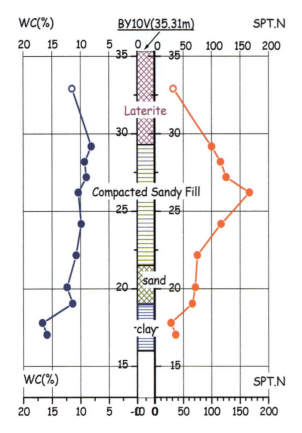
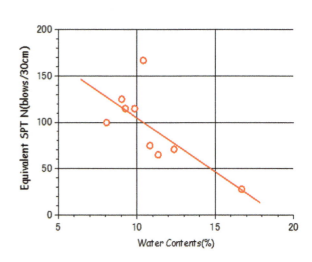

Figure 43. Water content and SPT N-values at BYV2010.   Figure 44. Water content vs. SPT N-values.

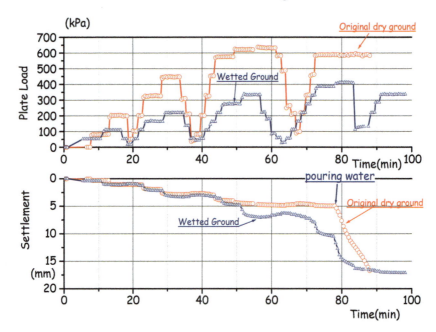

Figure 45. Charts of load time (upper) and settlement time (lower) for plate load tests on normal (in red) and wetted (in blue) original fill, Bayon.

Figure 43 shows the distribution of water content of the sampled soils, together with SPT N-values. The relationship between the water content and N-value is shown in Figure 44, and indicates a strong relationship between them.

Another example of the effects of water upon the sandy fill in Bayon is indicated by a plate load test, as shown in Figure 45. The diameter of the load plate was 15 cm. The load test was carried out in two stages.

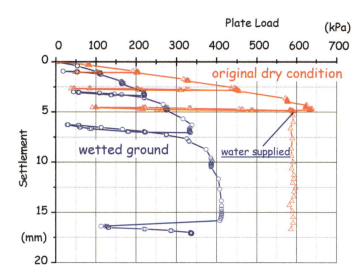

Figure 46. Results of load–settlement curve.

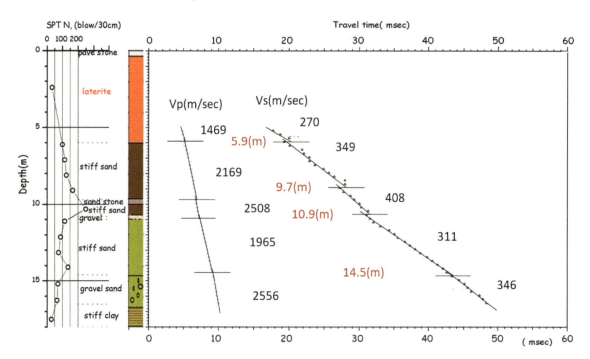

Figure 47. Results of P-S logging for the boring BYV2010.

The first stage was performed under normal water conditions, with a water content of 10.5%. When the load was considered to have been close to reaching the yielding step, water was poured into the ground. The second stage involved wet conditions for the same load plate to study the effects of water on the bearing capacity. The increase of water content resulted in significant decrease of the bearing capacity, from 650 to 400 kPa as shown in Figure 46. In an unsaturated state, the sandy compacted fill in Bayon shows very great strength due to strong negative pressure. However, it begins to be weakened under water infiltration because of the disappearing of the negative pressure that had attracted sand particles to one another.

7.6.4 P-S logging

P-S logging was performed to study the mechanical characteristics of the stiff sand (Akinaga *et al.*, 2015). The results are shown as the travel times of compression (P-) and shear (S-) waves in Figure 47 and are summarised in Table 1. The velocities of P- and S-waves, V_P and V_S, change with SPT N-values.

Table 1. The results of P-S velocity logging for the boring BYV2010.

Ground	Depth (m)	SPT (blow/0.3 m)	V_P (m/sec)	V_S (m/sec)
Laterite	0–6	32	1,469	270
Stiff sand	6–10	100–150	2,169	350
	10–12	150–250	2,508	410
	12–15	80–150	1,965	310
Natural gravel sand	15–17	70	2,556	360

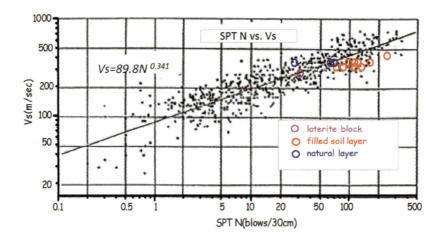

Figure 48. Plot of S-wave velocity (V_S) vs. SPT N-value at Bayon.

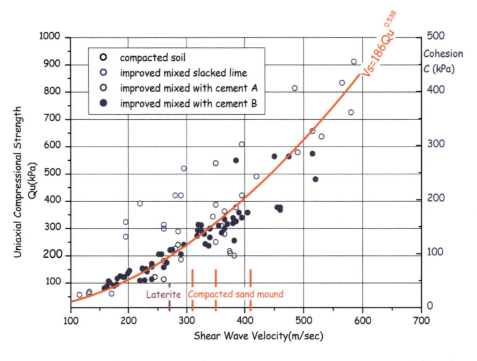

Figure 49. Relationship between shear wave velocity and uniaxial compression strength derived from experimental data.

The experimental relationship between V_S and SPT N-value is plotted in Figure 48 for the study in Bayon. It shows that the velocities of S-wave at Bayon fall within the range of previously established experimental relationships (Imai, 1982).

Figure 49 shows the relationship between shear wave velocity and uniaxial compression strength established in past experiments (Kita, 1982; Asaka, 2005).

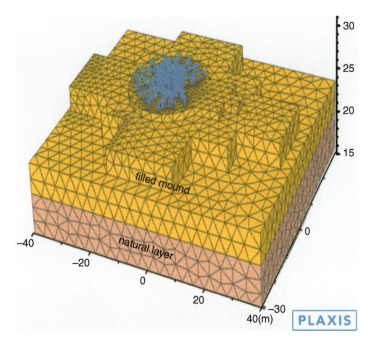

Figure 50. 3D modelling for FEM analysis.

8 GEOTECHNICAL STABILITY PROBLEM OF MAIN TOWER

The present state of the main tower of Bayon temple has been summarised in the previous section and has revealed possible questions relating to the stability of the foundation. One concerns the bearing stability of the foundation. Is there enough bearing capacity in the stone foundations to support the main tower? Another concerns the effects of the vertical shaft upon the foundation. In this section, the foundation mound is analysed with the Finite Element Method (FEM) to evaluate its stability as the base for the main tower.

8.1 *Modelling the foundation of the main tower of Bayon*

Because it is difficult to simulate the strength of the unsaturated sand by effective stress, total stress analysis using PLAXIS (Delft, The Netherlands) 3D FEM software was performed to understand the stability characteristics underlying the main tower of Bayon temple.

A 3D model for soil structure was constructed, as shown in Figure 50. The effects of the upper structures are considered as the vertical load distribution of the basements of the structures.

The upper structures consist of the central main tower and eight sub-towers, as shown in Figure 51.

The unit density of stone of $23\,kN/m^3$ was multiplied by the total volume of the tower and the estimated load is shown in Table 2. The base stones stand independently of each other and the contact load was estimated as 1,482 kPa for the main tower and 470 kPa for each sub-tower.

The material property to be used in the FEM analysis is the important but difficult step. As the preliminary analysis, the constitutive model was assumed to be a Mohr–Coulomb model of total stress analysis. Three different cohesions for the man-made platform mound were adapted from an experimental relationship of uniaxial compressional strength and shear wave velocity, as shown in Figure 49. The range of shear wave velocities derived from the *in situ* P-S logging was used as assumptions of the shear wave velocity for three models in the analysis.

In the field, it is found that the strength parameters are strongly affected by unsaturated conditions; the present analysis just assumed uniform properties of the filled soil with different strengths to find the effects on the stability of the vertical shaft. The hardening soil model was adapted and the rigidity G^* in Table 3 corresponds to V_S and is assumed under the reference condition of confining pressure $p_r = 100\,kPa$. Rigidity

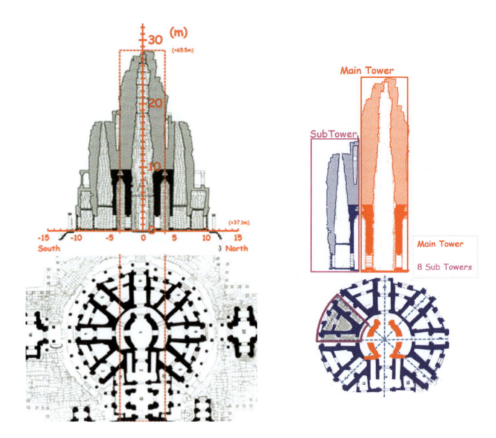

Figure 51. Structural shape.

Table 2. Estimation of load of the tower structure.

	Unit	Central tower	One sub-tower
Unit weight	kN/m^3	23	23
Volume	m^3	967	156
Total weight	kN	22,240	3,588
Contact area	m^2	15	7.64
Contact load	kPa	1,482	470

Table 3. Material properties in the FEM model.

Material		ρ_t (kN/m^3)	V_S (m/sec)	G^* (MPa)	ν	ϕ (degree)	C (kPa)
Platform mound	A	18.5	410	320	0.3	30	200
	B	18.5	350	230	0.3	30	150
	C	18.5	310	180	0.3	30	100
Natural ground		18.5	300	170	0.3	30	200

is assumed to depend upon the mean confining pressure p_m, based upon the following equation:

$$G = G^* \left(\frac{p_m}{p_r}\right)^{0.5}$$

where p_r is a reference mean stress.

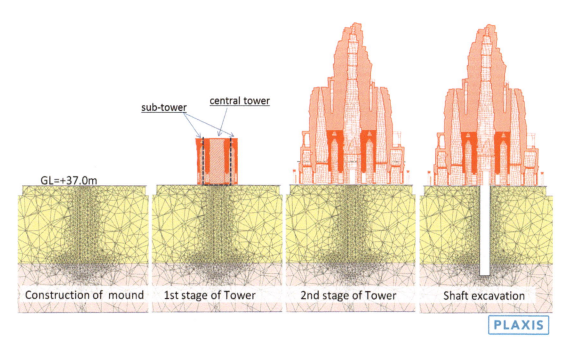

Figure 52. Loading steps of FEM analysis.

Table 4. Loading stages of construction and excavation of the vertical shaft.

	Initial step	1st step	2nd step	Shaft excavation
Main shaft		1000 kPa	1500 kPa	D = 2.7 m
Sub-tower		300 kPa	470 kPa	

Loading was divided into four steps following the construction of the filled mound, as shown in Figure 52 and Table 4.

As illustrated in Figure 52, after the construction of the platform mound, the lower part of the towers at the first stage and the remaining parts of the towers in the second stage are successively constructed. The last stage is the vertical excavation of a shaft 2.7 m in diameter to a depth of 14 m. The FEM was intended to find the strength of the soil mound necessary to maintain the stability of the foundation, rather than to explore the detailed behaviour of the foundation. One of the key points is the stability, or instability, associated with the excavated vertical shaft.

8.2 Result of FEM simulation

The FEM analysis did not intend to simulate the complete process of construction of the main tower but to obtain the basic status of the main tower in terms of foundation stability in relation to ongoing conservation of the main tower. The material properties assumed for the analysis have the lowest possible strength, with $C = 100$ kPa and $G = 180$ MPa (see Table 3). Figure 55 shows the distribution of the vertical stress beneath the foundation at the level of GL + 36.9 m for the two stages of the construction of the main tower and the excavation of the vertical shaft. The vertical stress increases from the first to the second stage of the construction. Most of the load is concentrated in the foundation of the main tower. The pattern of vertical stress following the main tower construction is not changed by the excavation of the vertical shaft, as illustrated in Figure 53. The vertical distribution of settlements of the ground surface are shown in Figure 54 in vertical section along a NW–SE orientation (indicated in Figure 53). The settlement is at a maximum at the foundation of the central tower where the maximum vertical stress is also found.

The excavation of the vertical shaft causes an increase in the shear stress around the shaft wall. The deviatoric stress on a vertical plane of NW–SE section is shown in Figure 55, which shows that deviatoric stress has

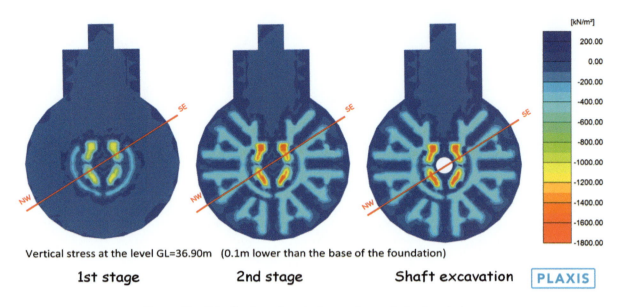

Figure 53. Distribution of vertical stress beneath the foundation.

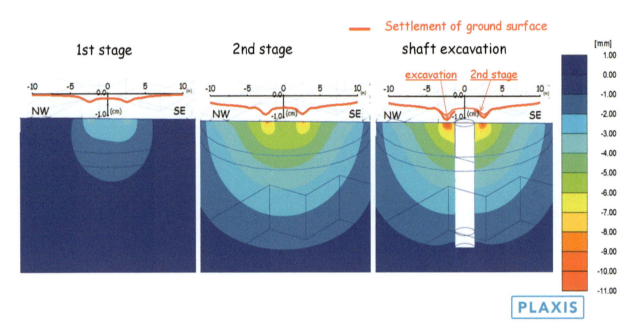

Figure 54. Ground surface settlement distribution in vertical section along a NW–SE axis.

increased and extended downwards around the vertical shaft. The increase of the deviatoric stress will cause yielding of soils around the shaft. Figure 56 shows the generation of a yielding zone around the foundation as well as the wall of the shaft.

Figure 56 also shows the yielding zone around the foundation before the shaft excavation. The increase of the yielding zones is also shown with the decrease of the cohesive strength of the material. In the model with $C = 200$ kPa, the shaft excavation does increase yielding around the shaft wall when compared with the state before excavation. However, in the model with $C = 100$ kPa, this increase in yielding is found to have expanded to the upper part of the shaft wall as well. Another yielding zone is identified at the bottom of the shaft where the natural sand has failed.

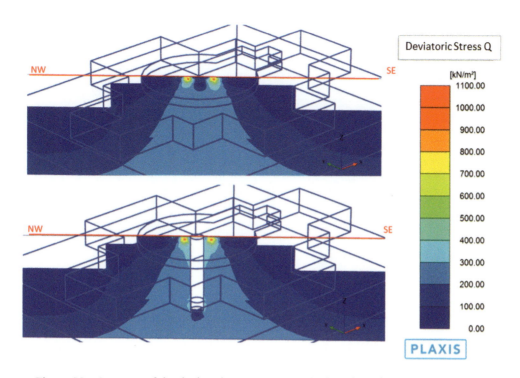

Figure 55. Increase of the deviatoric stress on a vertical section of NW–SE orientation.

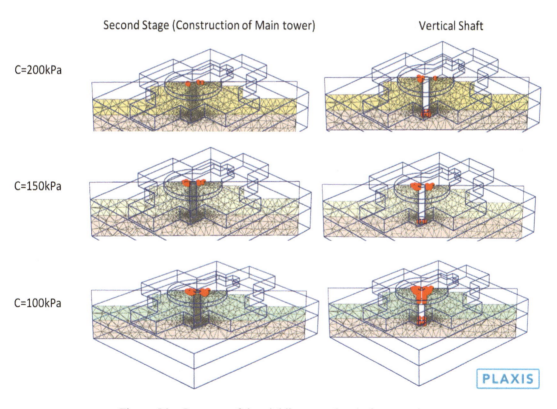

Figure 56. Increase of the yielding zone by shaft excavation.

8.3 *Safety of the present foundation mound*

The FEM analysis showed the basic behaviour of the main tower of Bayon on the foundation mound of uniform man-made fill. The result of the vertical distribution showed the decrease of the settlement outwards at sub-tower zone under the assumption of uniform ground.

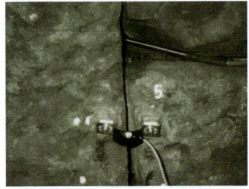
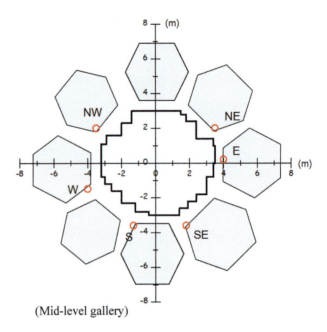

Figure 57. Gap sensor and temperature gauge.　　Figure 58. Installed monitoring points for sub-tower.

It became clear that the outwards inclination of the side-columns of the sub-tower is caused by ground. There should be some weakening ground at the peripheral zone. However, there is no deformation or displacement of the masonry stones of the foundation except for the opening of the gaps between stones that comes from the upper part of the stones.

The vertical shaft had first been excavated several hundred years ago and was rediscovered in 1933. Historical evidence indicates structural stability at present. Based upon the present situation of little deformation of the stones and ground near the vertical shaft, the cohesive strength is estimated as greater than $C = 150$ kPa.

9 PREVENTIVE CONSERVATION BASED UPON MONITORING

Currently, because of the uncertainty of ground characteristics, monitoring is a common practice during geotechnical construction to ensure the safety of construction. In conservation work on heritage structures, it is also recommended that a step-by-step approach is taken by monitoring the behaviour of the structure (ISO13822). To safeguard the Bayon temple, JASA has established basic weather monitoring as well as practising underground water level and extended monitoring of various other phenomena. Here, a typical case study of monitoring is introduced to identify the characteristic behaviour of the structures in Bayon temple.

9.1 Gap opening between stones

At the mid-level of the main tower, there is a gallery space between the central and sub-towers. Figure 57 illustrates a gap sensor and temperature gauge, which were installed at several points between stones of the sub-towers at the level of the mid-level gallery of the main tower, as shown in Figure 58. Figure 59 shows the changes of gaps between stones with hourly maximum wind velocity.

In Figure 59, the gap changes are more or less cyclical in nature with some trends. The crack in the stone wall may possibly have been caused by the widening of the gap between the stones. The crack is considered as the beginning of the first step of structural failure.

Figure 60 shows a portion of the monitoring records for the gap changes for three months from May to July 1998, based on an hourly sampling rate.

The wind velocity was also monitored at the top of the southern library in the Bayon temple. Figure 60 also shows hourly maximum wind velocity with time for the monitored data at the NW sub-tower.

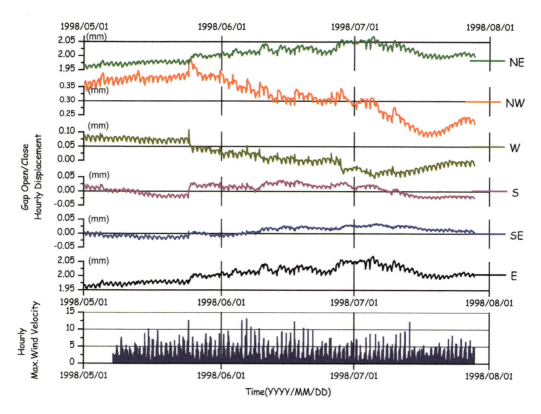

Figure 59. Change of distance in stone gaps vs. hourly maximum wind velocity.

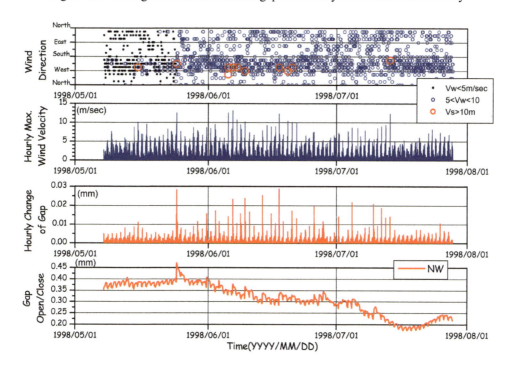

Figure 60. Hourly change of gap at NW monitoring point vs. wind velocity.

Gap changes with time are grouped into two types. The first type is temperature-dependent and operates on a daily cycle. Higher temperatures induce expansion of stone volume and result in decrease of the gaps. Gap changes depend upon changes of temperature and also upon the various different conditions of confinement of the stones. The second type of gap change does not follow a daily cycle but is spontaneous in nature, as seen, for example, in the significant changes around 23 May and 26 June visible in Figure 59 and the lowest

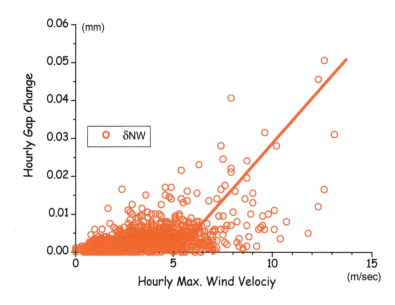

Figure 61. Hourly gap change vs. hourly maximum wind velocity.

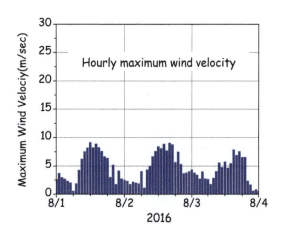

Figure 62. Hourly change of wind at Bayon.   Figure 63. Maximum wind velocities at Angkor.

graph of Figure 60. These changes happen not only at a single monitoring point but at several points at the same time as a common phenomenon. The hourly change of gap was calculated and is shown in Figure 60 for the NW monitoring point, together with distribution of wind direction for different velocities of the wind. Figure 61 shows the relationship between hourly gap change and hourly maximum wind velocity. It was found that strong wind velocities cause changes in the gaps between stones. The directions of winds stronger than 10 m/sec are plotted as red circles at the top of Figure 60. These strong winds are found to blow from a WSW direction, which is caused by the monsoon in the rainy season. The gap change remains within a range of about 0.01 mm for velocities of wind below 5 m/sec. The stronger wind velocities, above 5 m/sec, have caused large gap displacements. The gaps caused by strong winds can be further grouped into two different responses of reversible and non-reversible nature. Thus, the gap displacement on 23 May is found to be reversible at the NW monitoring point but non-reversible at other points in the W, S and E. These non-reversible gaps accumulate like the NW point does after 24 May.

When the reversible mechanism remains in the structure, these accumulated gaps may be recovered and remain within a balanced range. However, the reversible mechanism is not necessarily always guaranteed at every gap forever. The distribution of hourly maximum wind velocity over a period of three days at the start of August 2016 is shown in Figure 62 and the maximum wind velocity observed in Angkor in the past (Weather2, 2016) is shown in Figure 63.

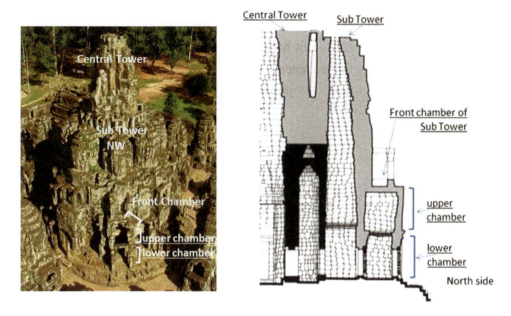

Figure 64. Front chamber of the NW sub-tower.

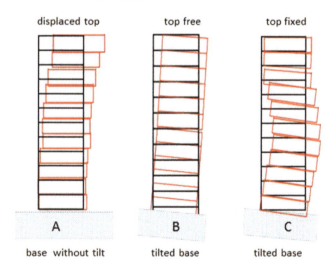

Table 5. Tilt of base and deformation.

	A	B	C
Top	Displaced	Free	Confined
Base	No tilt	Tilted	Tilted

Figure 65. Three modes of deformation.

As shown in Figure 62, stronger winds are observed in the daytime and the maximum wind velocity is usually below 10 m/sec. However, based upon the past records from the Siem Reap Weather station in Angkor (Figure 63), the maximum wind velocity can be more than 25 m/sec. These strong winds are believed to have caused irreversible displacement of the structure, described in the next section.

9.2 Monitoring of mechanism of inclination and gap opening

The gap opening in the wall of the NW sub-tower was shown in Figure 22 and was discussed in relation to the inclination of the wall. Figure 64 shows a photographic view and section of this wall.

The principal characteristic of the gap opening in the front chamber of the NW sub-tower is the increase of the gap with the height of the stone of the wall. Three possible kinds of mechanism to produce this mode of displacement are listed in Table 5 and shown in Figure 65.

The actual mechanism could be identified by monitoring the change of the gaps as well as the tilting of the foundation stones. These monitoring results will provide valuable information for identifying the mechanism of the deformation and the most appropriate approach for safeguarding the stone monuments.

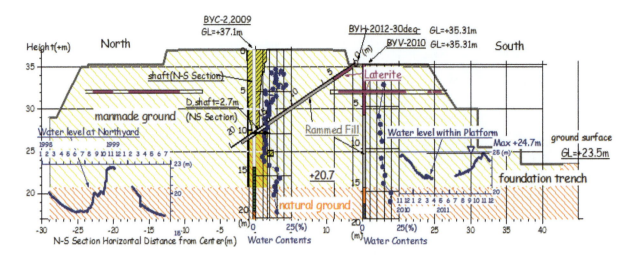

Figure 66. Water content in soil mound and seasonal change of underground water.

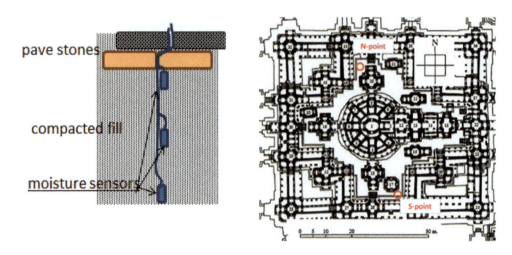

Figure 67. Installation of moisture sensors and location.

9.3 *Water condition in platform mound*

As already discussed, the foundation of the filled mound consists of very stiff sand whose strength is derived from the unsaturated condition of water in the soil. Water condition is the important actor in keeping the foundation of the main tower stable. The seasonal changes in the underground water outside the temple in borehole BY1994 and inside the mound are shown in Figure 66. The highest level of underground water is reached for both locations in November, at the end of the rainy season.

The lowest water levels are monitored in May at the end of the dry season. The water levels in the mound are always higher than those outside the temple by a few metres. The water contents of soil sampled from boreholes BYC2009 and BYV2010 were measured and are plotted in Figure 68. The water contents near the surface were 15–20% and decreased with depth to 5–10% with the minimum at GL − 5 m to GL − 10 m. Water in the foundation mound is provided mainly from rainfall that infiltrates the mound from the surface of the terrace.

9.4 *Water seepage from ground surface*

Rainwater is the main source for the increase of the water content in the platform mound. JASA installed moisture sensors at several depths from the platform surface at points on the mound as shown in Figure 67 (Koyama *et al.*, 2015).

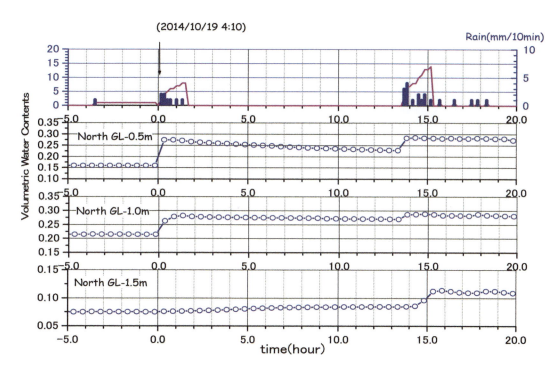

Figure 68. Response of moisture sensors under the surface of the mound at different depths.

Table 6. Response of sensor at GL − 1.0 m to rainfall of short duration.

Rain (mm)	Initial response	Max. level
0 < rain < 7	No response	No response
7 < rain < 17	No sharp response	Gradual rise to some maximum level
17 < rain < 20	Initial response is recognisable with delay of 0.5 hours	Delay of 2–3 hours to reach the maximum level
20 < rain	Delay is 0.5 hours	Delay is 2 hours

Moisture sensors were installed in the ground below pavement stones. The location of the moisture sensor installations is shown in Figure 67, and they were installed at three depths: GL − 0.5 m, GL − 1.0 m and GL − 1.5 m. Figure 68 shows the differing responses of the sensors at different depths to various amounts of rainfall. The sensor at the deepest position responds with the longest time delay. The time delay is about 0.5 hour for the depth of 0.5 m, which results in a seepage velocity of 1 m/hour. Two rainfall events were observed. In the first event, the sensor at the lowest depth of GL − 1.5 m shows little response to the rainfall. However, it did respond to the second rain event. The difference is the amount of the rainfall; the first event involved 8 mm and the second event 13 mm. Table 6 summarises the results monitored for common, short rainfall events of less than about three hours' duration. We may conclude that if the rainfall is less than about 20 mm, the rainwater could not infiltrate down to GL − 1.5 m but is retained within the soil in the upper surface and gradually evaporates. When the rainfall event exceeds 20 mm, the rainwater infiltrates down to GL − 1.5 m within two hours.

9.5 *Monitoring of the deformation of the tower by GPS*

In addition to the monitoring of gap displacements and the inclination of the structures, three Global Positioning System (GPS) sensors have been installed at the tops of the main tower and two sub-towers, as shown in Figure 69 (Masunari *et al.*, 2015). A differential GPS monitoring system was established in 2014 and the results of 3D displacements have since been monitored, as shown in Figure 70.

The GPS00 base station is located at the south side of the outer gallery of the temple. The GPS signals are monitored every 30 seconds and show some scattering. However, the displacements monitored are stable

Figure 69. GPS installations.

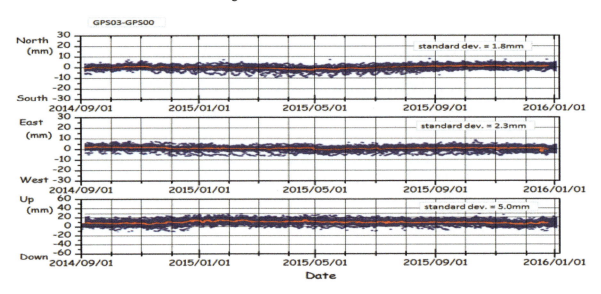

Figure 70. Displacement of top of the central tower according to GPS.

with scatterings of standard deviation 2–3 mm for horizontal, and 5–8 mm for vertical, components, as shown in Table 7. The monitoring of displacements of the tower by GPS is useful to understand the characteristic movement of the structure and provides continuous information on the stability of the main tower.

10 COUNTERMEASURES FOR CONSERVATION OF THE MAIN TOWER, BAYON

10.1 Results of assessment of structural safety

At present, three structural problems have been identified for the main tower of Bayon temple. The first one is the collapse of stones at the outer periphery of the tower, which has resulted in the loss of symmetry of the

Table 7. Standard deviation (mm) of the displacements monitored by GPS at Bayon.

	GPS01-GPS00	GPS02-GPS00	GPS03-GPS00
North–South	2.1	3.0	1.8
East–West	2.9	2.1	2.3
Up–Down	6.7	7.5	5.0

tower as representing a lotus flower. The second problem is the outwards spreading displacement of stones of the tower, which has caused inclination of upright pillars and gaps between stones in walls. These instabilities of the masonry structure are considered to have been caused by the long-term actions of plant roots and strong winds. The third structural problem is the vertical shaft in the platform foundation. The excavated shaft has resulted in a weakening of the bearing capacity of the foundation. However, based upon the historical fact of having stood for several hundred years without any problems, it is reckoned that the platform mound with the excavated shaft has enough strength to support the main tower for the time being. The mound consists of filled sand that shows sufficient strength under dry conditions but will lose that strength under wet conditions.

10.2 Authenticity of foundation of Bayon temple

As indicated in 1.5.3 of ISO 13822 (ISO, 2010), the foundation is one of the key elements of a heritage structure and it is important to discuss its authenticity. The foundation system in Bayon temple shows the following unique geotechnical features or character-defining elements as indications of its authenticity:

1. The trenched foundation system with three-stepped man-made fill with well-compacted sand is very unique and the top terrace mound of 14 m in thickness is also unique in supporting the main masonry stone tower of 31 m height by direct shallow foundation in Bayon temple.
2. Another special feature is the material of the filled soil, which consists of compacted kaolin sand with uniform grain-size distribution and shows very high SPT N-values of 100–150. This is the result of strong negative pore water pressure under unsaturated water conditions.
3. It is also amazing that the vertical shaft, which was probably dug at the centre of the base of the foundation sometime in the 13th/14th century AD, has been stable for several hundred years. The shaft itself might well be regarded as having another character of authenticity.

10.3 Countermeasures for foundation in Bayon temple

The conservation problems of the foundation of the main tower in Bayon have come to focus on two points: keeping the mound stable in the face of the increasingly heavy rain anticipated in future due to global warming, and the stability of the vertical shaft. Countermeasures against these problems should be performed under the three basic principles of heritage conservation: minimal intervention, an incremental approach, and removal measures for the integrity of the authenticity of the heritage value. The importance of protecting the top terrace mound from heavy rain has become clear. It requires an effective method of providing an impervious layer beneath the pavement stone and the sandy soil foundation. The intervention must be removable and comparatively simple, and should be monitored to confirm its effects.

JASA is planning to install a replica of the original Buddha at the centre of the main chamber of the central tower. The base of the replica will be placed just above the vertical shaft. The shaft was backfilled with very loose soil and has no bearing capacity to support the replica. The foundation for the replica, which weighs about 15 tons or 150 kN, should be carefully designed to support the Buddha statue (Nakagawa et al., 2014). Based upon the principle of a step-by-step approach, JASA will take the first step of conservation work to provide countermeasures to prevent rainfall infiltrating the ground and will maintain preventive monitoring of various behaviours of the ground and structure. The possible countermeasures involved in this first step are listed in Table 8.

Table 8. Comparisons of possible countermeasures to avoid an increase of water in the foundation.

Method	Countermeasure for foundation of main tower		
	Shelter roof	Surface grouting	Impervious layer
Removal measure	Yes	Difficult	Yes
Incremental approach	Yes	Yes	Yes
Minimal change		—	
Technical/cost	Expensive		
Comment	Not good appearance. Only acceptable for the top of the tower.		Monitoring system for moisture in the ground shall be established before and during the countermeasure work.

11 CONCLUSIONS

The problems of the main tower of Bayon at the Angkor monument have been described. The main tower is made of complex stone masonry in a combination of the central tower and eight sub-towers. The non-symmetrical shape of the top of the main tower was caused by collapse of the stone columns from which it was constructed. Outwards inclinations of the stone pillars of the sub-towers, and upwards increases in the vertical gaps in the stone walls are major deformations; they are estimated not to have been caused by the base ground but rather by the outwards splitting of the stones of the upper structures. Based upon current practice in geotechnical engineering, it is amazing to discover that a direct foundation on man-made fill is able to support the main tower. This secret was found and identified as due to the peculiar characteristics of the sandy soil. The kaolin sand, which shows very strong strength in unsaturated conditions, is the key to support of the heavy stone structure and is identified as a character-defining element of the authenticity of the foundation of Bayon temple. The preventive observational procedure is the most appropriate method for the conservation of the main tower at Bayon temple. Other masonry structures close to nearby ponds show differential settlements. Worldwide, heavier and more irregular rainfall is anticipated in the future and preventive monitoring is one of the possible countermeasures in preventing unexpected failures in heritage structures.

ACKNOWLEDGEMENTS

The authors express their sincere thanks to Prof. T. Nakagawa, JASA team leader for his understanding of the importance of geotechnical engineering for the safeguarding of Angkor, Prof. I. Shimoda, Prof. M. Araya, Prof. N. Yamamoto, Prof. K. Nakagawa, Prof. T. Adachi, Dr T. Masunari, Prof. T. Koyama, Prof. T. Tokunaga, Prof. K. Mogi, Prof. S. Yamada, Dr Y. Akazawa, Dr M. Ishizuka for the dialogues on ideas on soil, structure and their interactive behaviour, Prof. T. Haraguchi for his providing photos from air by a drone, Mr A. Kitamura and Mr Y. Ide for their excellent boring technique, Mr Y. Akinaga, Mr K. Yamada, and Mr H. Yokota for their cooperation in PS-wave logging, Mr H. Ito for his vital advice in the use of the FEM program PLAXIS, and Mr R. McCarthy for his useful suggestions on local soils, including English expressions.

REFERENCES

Akinaga, Y., Yamada, K., Yokota, H. & Iwasaki, Y. (2015). *PS wave logging,* Annual technical report on the survey of Angkor monument 2012–2013, JASA, Tokyo. pp. 138–142.

Asaka, Y., Abe, T., Katsura, Y., Sugimoto, H. & Tatsumi, Y. (2005). *Non-Destructive Technique Using Shear Wave Velocity to Inspect the Quality of Cement-Treated Ground*, Technical Research Report, Shimzu Institute of Technology, Vol. 82, pp. 23–29 (in Japanese).

Imai, T. & Tonouchi, K. (1982). Relationship between SPT, N and S-wave velocity and cases of the application, Foundation Engineering, Tokyo, Vol. 16, No. 6, pp. 70–76 (in Japanese).

ISO. (2010). ISO13822 Annex I Heritage Structures, ISO13822:2010 Bases for design of structures – Assessment of existing structures. Geneva, Switzerland: International Organization for Standardization. http://www.iso.org/iso/catalogue_detail?csnumber=46556

Iwasaki, Y., Fukuda, M., Haraguchi, T., Kitamura, A., Ide, Y., Tokunaga, T. & Mogi, K. (2014). *Structure of platform mound of central tower based upon boring information*, Annual technical report on the survey of Angkor monument 2012–2013, JASA, Tokyo. pp. 93–113.

Kita, D., Kubo, Kubo, H., & Urushibara, K. (1982). Estimating Strength of Stabilized Soil by Velocity of Elastic Waves, Research Report of Ohbayashi Construction Company, Vol. No. 25.

Koyama, T., Yamada, S., Iwasaki, Y., Fukuda, M. & Shimoda, I. (2015). *Installation of moisture sensor*, Annual technical report on survey of Angkor monument 2014–2015, JASA, Tokyo. pp. 132–137.

Masunari, T., Iwasaki, Y. & Ishizuka, M. (2015). *GPS monitoring system for central tower, Bayon*, Annual technical report on survey of Angkor monument 2014–2015, JASA, Tokyo. pp. 111–114.

Nakagawa, T., Shimoda, I. & Ishizuka, M. (2014). *Bayon Great Buddha Project: Restoration, reconstitution, and reinstallation of the original Bayon Buddha image,* Annual technical report on the survey of Angkor monument 2012–2013, JASA, Tokyo. pp. 163–164.

Narita, T., Nishimoto, S., Shimizu, N. & Akazawa, Y. (2000). *Outline of excavations and investigation at the outer gallery of Bayon complex*, Annual technical report 2000, JASA, Tokyo. pp. 3–22.

Shimoda, I., Yamamoto, N., Iwasaki, Y. & Fukuda, M. (2009). *Excavation survey of the central tower chamber*, Annual technical report on survey of Angkor monument 2008, JASA, Tokyo. pp. 67–88.

Weather2. (2016). http://www.myweather2.com/City-Town/Cambodia/Siem-Reap/climate-profile.aspx

Geotechnics and Heritage: Historic Towers – Lancellotta, Flora & Viggiani
© 2018 Taylor & Francis Group, London, ISBN 978-1-138-03272-9

Safeguarding the leaning Minaret of Jam (Afghanistan) in a conflict scenario: State of the art and further needs

A. Bruno
StudioBruno, Via Asti 17, Torino, Italy and UNESCO Expert

C. Margottini
ISPRA, Geological Survey of Italy, Rome, Italy and UNESCO Chair at Florence University

L. Orlando
Department Civil and Environmental Engineering, University "La Sapienza", Rome, Italy

D. Spizzichino
ISPRA, Geological Survey of Italy, Rome, Italy and UNESCO Chair at Florence University

ABSTRACT: The Minaret of Jam is located in the centre of Afghanistan, some 200 km east of Herat at 1,904 m above sea level and at the confluence of the river Hari Rud with its tributary the Jam Rud. The minaret was probably built between 1163 and 1203 and is composed of four tapering cylindrical shafts. In 2002, UNESCO declared Jam Afghanistan's first World Heritage Site. The minaret, 65 m high, is presently affected by an important change of verticality, of about 3.4°. The reason for this lean is not very clear but is likely to be due to the undercutting excavation of the foundation produced by the nearby Hari Rud and its related flood effects. In order to develop a long-term conservation plan, seismic, geological, geomorphological, geotechnical and geophysical investigations were performed between 2003 and 2005. Some initial restoration works were also performed in the period 2005–2009, to reduce the structural degradation of the masonry and to enhance protection from the river, specifically after a powerful flood in 2007. Based on the available information, knowledge of the site and the professional expertise of the authors, some short-term emergency interventions have also been proposed. Future investigation and restoration works have been recommended and these should involve proper field investigations and monitoring. In conflict areas such as this, where it is essential to preserve the heritage from disappearance, there is not the time nor the technology for consolidated procedures. In these situations, recourse to the professional judgement of the various experts involved is the main operational approach, as has been demonstrated in many UNESCO case studies. This is also the case for the Jam minaret, with hope that the recent peace in Afghanistan will also allow the shift from emergency interventions to medium- and long-term planning.

1 INTRODUCTION

The leaning Minaret of Jam, one of the tallest in the world, was declared Afghanistan's first World Heritage Site by UNESCO in 2002. The minaret is located in the centre of Afghanistan, 215 km east of Herat and 1,904 m above sea level, at the confluence of the River Hari Rud and its tributary the Jam Rud, where three valleys converge, surrounded by mountain ranges that reach a maximum height of 2,300 m.

The Minaret of Jam rises to a height of 60.41 m from ground level and is composed of four tapering cylindrical shafts. The tower structure, constructed with bricks and lime mortar, is composed of two parts: the external one, made of four tapered truncated blocks with ring-like cross sections, one upon the other, and the internal one, composed of a tapered block with circular cross section, which ends at a height of 43.50 m. Two spiral staircases lie between the external and the internal parts, and also join them. The actual position of the basement is not known in detail, but from the existing surveys we can see that the entrance to the minaret is buried about 4 m below the present topographic surface and alluvial deposit. So the total height should be about 64.5 m, without considering the original ground floor and foundation. The minaret is currently leaning at about 3.4°, the reason for which has been investigated by UNESCO missions since 2005.

The present chapter reports the investigation executed in the Jam valley to understand the causes of the leaning of the minaret and the emergency intervention required after the 2007 flood. The investigations were performed in a time period when access and logistics to the Jam area were considered safe for international experts. After 2006, these conditions were no longer fulfilled and all the local activities were conducted with the great support of the Afghan Minister of Information and Culture, with training activities and capacity building in Kabul, that is, for the emergency intervention after the 2007 flood.

2 THE HISTORY FROM INSIDE

The Minaret of Jam (Figure 1) was probably built between 1163 and 1203 in the reign of the Ghurid sovereign Ghiyas-ud-Din. Its emplacement probably marks the site of the ancient city of Firuzkuh, believed to have been the summer capital of the Ghurid dynasty. It is likely that it is the only monument to escape the destruction wreaked by Genghis Khan, and it remained unscathed and largely unknown for over seven centuries. The minaret itself and the archaeological remains surrounding it offer extraordinary testimony to the grandeur and refinement of the Ghurid civilisation, which dominated the region in the twelfth and thirteenth centuries. Besides being an exceptional artistic testimony to twelfth-century Islamic culture, the Minaret of Jam also provides historical and political evidence about this isolated region near the most important commercial thoroughfare in antiquity, linking Constantinople to India via Afghanistan, Herat and Kabul.

The wider discovery of this extraordinary monument occurred quite recently, because the first official mention of its existence dates back only to May 1944, when Ahmed Ali Koazad, president of the Afghan History Society, published an article in the journal *Anis* relating a conversation he had had with His Excellency Abdullah Khan Malekyar, the Governor of Herat, who mentioned the Minaret of Jam to him, having been the first to discover and photograph it before 1943.

After its rediscovery in 1944, the minaret was again 'lost from view'. Many expeditions were mounted to locate it in the ensuing years, but were unsuccessful due to the immensity of the territory and the difficult terrain. It was not until 19 August 1957 that the French archaeologist André Maricq finally found the minaret and published his discovery in the *Mémoires de la Délégation Française en Afghanistan* in 1959 (Maricq & Wiet, 1959). In 1960, two envoys from DAFA (Délégation Archéologique Française en Afghanistan, or the French Archaeological Delegation in Afghanistan), Le Berre and his assistant Marchal, photographed the decorative panels and conducted an initial and very brief study of its architectural structure, uncovering the entrance with its two helicoidal stairways about 4 m below the current level.

The octagonal brickwork structure, bonded with cement, is about 60.41 m high (from the current ground level) and comprises three cylindrical 'shafts'. The original entrance lies about 4 m below the present topographic surface, below a layer of alluvial deposits, and it is currently inaccessible. Access to the minaret is via a narrow window giving on to one of the two helicoidal stairways that lead to the top of the first 'shaft', about 40 m up. The outside of the minaret is entirely covered in complex monochrome decoration in haut- and bas-relief. A special technique has been used to incorporate complex geometric designs, including quotations from verses of the Koran in Kufic and Nashi scripts. The stylised inscriptions are a striking mix of vertical bands and circular tracings recalling a geometric floral design. Halfway up the minaret, a band of turquoise glazed tiling lends a unique touch of colour to the uniform brickwork surface.

The Minaret of Jam provides proof of the long architectural evolution of minarets in general. Eleventh-century minarets were circular (as in the Islamic Republic of Iran, at Isfahan and Barsiyan), but in Jam the base is an octagonal structure of 9.05 m (according to the latest surveys conducted in October 2002), from which a 40 m high conical trunk emerges, the diameter decreasing nearer the summit (ibid.).

2.1 *Risk of collapse*

Aware of the monument's importance and the lack of precise data regarding its conservation status, at the beginning of the 1960s the Afghan Government commissioned ISMEO (Istituto Italiano per il Medio ed Estremo Oriente, or the Italian Institute for Middle and Far East Studies, based in Rome), which was already

Figure 1. The Minaret of Jam (photo © C. Margottini, 2004).

working in Afghanistan, to undertake some research. Its task was to collate as much information as possible in order to draw up a programme to safeguard the minaret. In September 1961, Andrea Bruno was commissioned by ISMEO to carry out the first survey of the minaret and put forward a protection and restoration programme that included consolidation of its base (Figure 2).

At that time, conservation of the monument was seriously jeopardised given the risk of collapse from river erosion of the base, causing it to list on its axis (Figure 3). As the projected action required an examination and preliminary analysis of a considerable amount of data, it could not be undertaken hastily, especially bearing in mind the difficulties of reaching the site.

The minaret was probably built at the height of the Ghurid dynasty, even though no date is mentioned in the inscriptions. The name of Ghiyath al-Din appears three times and, according to certain historical texts, there was apparently a mosque there in 1192–93, which might have been destroyed on the orders of Genghis Khan in 1221 or carried away by a river spate. The recent mission involving a geologist (August 2002) uncovered evidence that might confirm this latter hypothesis in the form of the 4 m layer of alluvial deposits covering the base of the minaret. This finding should be studied carefully by future missions. Decisions regarding the best way to restore the minaret cannot be taken until a thorough survey has been conducted.

To prevent other threats to the stability of the minaret, whose base is threatened by river erosion, traditionally built protective gabions of stones and tree trunks were set in place with the aid of the inhabitants of the neighbouring village in August 1963 (Figure 4).

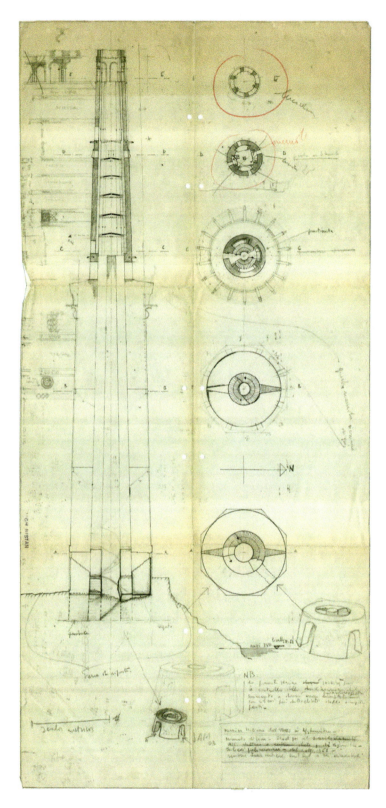

Figure 2. First survey of Jam minaret in 1961 (© A. Bruno).

In June of the following year, a brick and mortar sampling programme was conducted, but it was impossible to carry out surveys of the subsoil due to the complexity of the problems that had to be dealt with, including lack of machinery and specialised personnel.

Many expeditions were undertaken at the end of the 1960s to reach the minaret from the village of Jam along the Jam Rud, and a wooden bridge was built across the Hari Rud to access the Bedam Valley.

Figure 3. The Minaret of Jam (photo © A. Bruno, 1961).

Figure 4. The Jam minaret in 1961: (a) with the Hari Rud close to the structure and the first gabions in 1963; (b) view from the top (photos © A. Bruno, 1961, 1963).

In 1974, at the request of the Afghan Government and within the framework of the UNESCO-UNDP Project (Restorations of Monuments in Herat – Strengthening Government's Capability for the Preservation of Historical Monuments), UNESCO commissioned Andrea Bruno to undertake a thorough restoration study of the minaret because he was already in Herat at the time. Only then, thanks to the opening of a road suitable for vehicles, was it possible to plan a restoration project and finally transport the equipment and materials needed to the foot of the minaret. A small house was built alongside the monument to accommodate the labour force, as well as a warehouse for the building materials.

2.2 *A break in the project*

Between 1975 and 1980, when the citadel and minarets in Herat were being restored and a team of Afghan architects and engineers specialising in restoration work was set up, UNESCO financed an emergency intervention in Jam as well (in 1978) that was focused on the installation of protective metal gabions imported by sea and road from Europe, a step that was to be the first phase of the project. The gabions were to be filled with stones from the site to protect the base of the minaret against river erosion, but they were also meant to facilitate surveys and digs in order to obtain better archaeological information about the site before starting work to reinforce the base and foundations.

By September 1979, stones had been accumulated near the minaret. This initial operation, as well as the testing of the minaret's foundations and the clearing away of the debris around the base, was to be completed the following year. Shortly afterwards, however, when the conditions required for action had finally been met, war broke out and made any further intervention impossible.

During the Soviet occupation, while UNESCO continually expressed its interest in conserving Afghan cultural heritage and, in particular, the city of Kabul and its archaeological museum, it was never able to take effective action. The question of safeguarding the monument was delegated to non-governmental organisations, especially the Society for the Preservation of Afghanistan's Cultural Heritage (SPACH), an organisation based in Peshawar, which for all those years did very useful work involving documentation and emergency actions. It continually drew attention to the problems of conservation in dramatic conditions in which the priorities were of a quite different nature.

2.3 *Contact renewed*

It was not until 1999 that the minaret received any serious attention and a new mission was carried out following a report from SPACH highlighting the dangers that threatened it. Under Taliban rule, 20 years after his last visit, Andrea Bruno was able to reach the minaret again on 2 August 1999, together with an Afghan engineer, thanks to a special agreement between the warring parties – Mujahidin and Taliban – who were entrenched on either side of the Hari Rud.

Although that inspection visit lasted barely an hour, after which the experts were forced to leave in order to avoid security problems that might have compromised the mission's success, it achieved its ultimate objective, which was to check the state of the minaret and, as far as possible, compare it with the situation in 1979. In that short time, a decision was taken on the kind of emergency action that was feasible under the circumstances. Those 20 or so years that had seen a series of tangible interventions making it possible to prepare for the protection and restoration of the minaret and to organise archaeological research throughout the area, had been followed by 20 more years (1979–99) of senseless destruction. Thus, by 1999 all the achievements of 1963–79 had been wiped out.

The last 20 years of 'holding one's breath' had led to a particularly difficult situation and shattered all hopes of swiftly ensuring effective protection for the minaret. The bridge, base camp and infrastructures which should have served as a worksite depot were completely destroyed; the metal gabions provided by UNESCO in 1978 had disappeared, as had the protective materials installed in the 1960s; only the stones that had been gathered to fill in the gabions remained.

We had, in other words, reverted to the situation that prevailed prior to 1960, one of complete isolation from the civilised world. After 30 years of care and effort, work had to begin again from scratch, and, what is more, in a situation of armed conflict.

During that brief 1999 mission, in agreement with the Afghan engineer, who made himself available to supervise the work, a decision was taken to reuse the remaining stones and, in the absence of metal gabions, to use tree trunks to provide the minaret with temporary protection. As in 1961, the necessary labour was recruited from the area around the village of Jam. However, the idea of reconstructing the house and bridge as in the 1970s was ruled out lest they serve as military positions against opposition forces (those of Commander Massoud).

In 1999 and 2001, SPACH and HAFO (Helping Afghan Farmers Organization) carried out protective measures in two phases. In the absence of gabions, the first phase involved an effective use of wood and stone (they successfully withstood a flood on 14 April 2002), while the second phase involved the use of metal gabions along the bank of the Hari Rud.

2.4 On course again

In March 2001, after the destruction of the Bamiyan Buddhas and the fall of the Taliban regime, UNESCO was again able to take over direct control of the situation and send a consultant to the site in a UN helicopter. Although the visit confirmed that the measures taken to protect the minaret's base had proved successful, it also revealed the disastrous results of illicit excavations that would seriously jeopardise the scientific results anticipated from the projected archaeological digs. Judging by the scale of the clandestine digs, those responsible for them must have made a huge profit. What precisely has disappeared cannot yet be determined. A project to build a road and a bridge also presents a threat to the archaeological site as it might destabilise the minaret and would ruin the archaeological zone.

The gravity of the situation justified urgent action to safeguard the monument. This was the strongest argument put forward during the International Seminar on the Rehabilitation of Afghanistan's Cultural Heritage, organised by UNESCO and the Ministry of Information and Afghan Culture in Kabul from 27 to 29 May 2002.

Meanwhile, the UNESCO World Heritage Centre and the Afghan authorities developed a nomination dossier, including important information obtained from the March 2002 mission report. The Government of Afghanistan showed exemplary commitment and dedication in this endeavour. It nominated the site for inscription on both the World Heritage List and the List of World Heritage in Danger for consideration by the International Council on Monuments and Sites and the World Heritage Committee at their 26th session in Budapest (24–29 June 2002). This led to the first inscription of an Afghan property on the World Heritage List. Recognising the urgent need to mobilise international technical and financial support to protect the property, the Committee accepted the Afghan government's wishes to immediately inscribe the site on the List of World Heritage in Danger.

In the wake of the serious destruction wrought in the Bamiyan Valley (particularly that of the giant statues of Buddha) and the partial destruction of Kabul Museum, the Minaret of Jam has become a symbol. There is, consequently, room for cautious optimism concerning the future of Afghanistan's cultural heritage.

Inclusion of the minaret on the World Heritage List has allowed UNESCO to initiate protection work by setting up a working group entrusted with the task of ensuring the monument's preservation and enhancement.

In August 2002, Andrea Bruno headed a mission that defined the basis necessary for an accurate survey of the area, the local water courses and a more complete study of the foundations with the cooperation of a specialised geologist (Bruno, 2002).

Following upon that mission, in October of that same year UNESCO commissioned a team with state-of-the-art equipment to carry out the second survey of the minaret to be conducted in the last 40 years (the only previous survey being the one undertaken by Andrea Bruno in 1961). That mission yielded a very positive result, providing UNESCO with plans of the minaret and with various precise measurements of its height, the cross sections of the walls, the exact incline of its axis and the site topography.

Those survey results and, above all, the measurements of the exact incline, are essential data for those responsible for planning consolidation work. Professor Macchi, an engineer specialising in the structural problems of such buildings as the Tower of Pisa and the bell tower of the basilica of St Mark's in Venice, has been commissioned to prepare a consolidation project for the minaret. The first part of the project consists of a series of soundings of the foundations by means of core-drill boring and a structural study to be conducted by specialised companies. A first preliminary investigation has been conducted by Professor Macchi and by a geologist, in order to decide the execution modalities of the soundings and the basis of the minaret. The results of those preliminary surveys have been elaborated (Bruno, 2003a, 2003b, 2004).

Figure 5. Meeting with local authorities during 2004 mission (photo © C. Margottini).

With regard to the archaeological mission, a preliminary reconnaissance has been conducted by Italian archaeologists from ISIAO (Istituto Italiano per l'Africa e l'Oriente, or the Italian Institute for African and Oriental Studies, formerly ISMEO), under the supervision of Prof. Giovanni Verardi, already operating in Ghaznì. The archaeological research will continue with a thorough examination of the clandestine digs to obtain enough information to allow Afghan archaeologists to excavate in a scientific fashion.

These initial actions will precede the main restoration project, which will involve other specialists and all necessary materials and machinery.

A new series of missions were conducted in 2004 (see Figure 5), 2005 and 2007 (with another short mission in 2014), allowing the generation of the data and the accomplishment of the works described in this chapter.

3 GEOLOGICAL CONTEXT

The Minaret of Jam is located in a narrow valley, at the confluence of two rivers: the Hari Rud and the Jam Rud. The local geology is strongly affected by its mountain environment, with a bedrock covered by coarse alluvial and talus slope deposits. In detail, the bedrock is composed of massive granite/granodiorite, covered by a few orders of alluvial deposit; the flanks of the valley are covered by debris cone/talus slope materials. A geological map of the area is depicted in Figure 6.

The alluvial deposit is characterised by a few orders of sediments, related to different alluvial stages. The most recent is still under formation in both valleys. The main part of the floodplain is composed of materials deposited since the 12th century, because they cover many archaeological remains, as detected by geophysical surveying. Even the entrance of the minaret is about 4 m below the present-day topographic level. This is the evidence for one or more important flash floods, burying the ancient town. As confirmation, large blocks have been found during the recent excavation by Andrea Bruno. It is not clear if the flash flood(s) was the main cause of destruction of the town or simply if it (they) occurred after destruction perpetrated by Genghis Khan, as mentioned in some historical records.

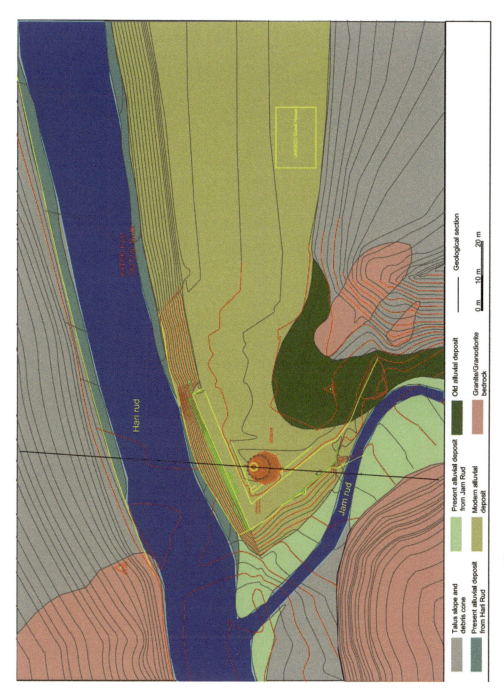

Figure 6. Geological map of the Jam minaret area.

The alluvial material is mainly composed of sand and gravel, etheroclastic and etherometric with an abundant presence of sand matrix. The grain-size distribution of the matrix (<2 mm in diameter) highlights the presence of sand (55%), silt (26%) and clay (14%), as shown in Figure 7.

Large parts of the flank are covered by talus slope/debris cone deposits, related to minor rock fall and debris accumulation from which the occurrence of rock avalanches and debris flow.

4 SEISMICITY OF THE REGION

There is no direct evidence of large earthquakes having occurred at the site in the historical and technical literature and the nearest epicentre lies about 170 km from Jam (Menon *et al.*, 2004). Historical and instrumental

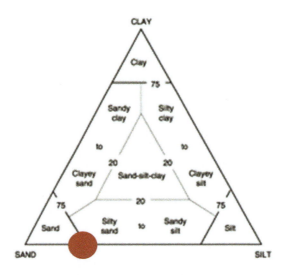

Figure 7. Sheppard sedimentological diagram for the alluvial deposit, restricted to the matrix with diameter <2 mm.

catalogues of earthquakes (Menon & Lai, 2004; Ambraseys & Bilham, 2014), from 25 AD to the present day, validate the above statement. However, according to Ambraseys and Bilham (2014), the historical record of Afghanistan's earthquakes is far from being complete in certain remote areas of the country. Fairly recent works have classified this region as a weak seismicity zone (Abdullah, 1981, 1993). Jam lies in close proximity to the Herat fault, a prominent right lateral strike-slip lineament running through north Afghanistan for about 1,100 km. For most of its length, the fault has not been associated in recent history with the occurrence of earthquakes.

Preliminary geological investigations at Jam by Borgia (2002) for UNESCO have revealed recent tectonic uplift in the region. Therefore, earthquakes of moderate intensity cannot be totally ruled out. Residents of the neighbouring village have experienced earthquakes, but the data is far too inadequate to determine the intensity of ground motion or the recurrence interval (Borgia, 2002).

The impact of earthquakes in the evolution of the minaret cannot be excluded but there is not clear evidence to associate its leaning with seismic forces. However, it is likely that seismic activity is an important threat to the long-term conservation of the minaret (Menon & Lai, 2004). Figure 8 documents the seismicity in and near Afghanistan from 25 AD to the present.

The future seismic hazard at Jam (Menon et al., 2004), deriving from a Probabilistic Seismic Hazard Assessment (PSHA), is low with an expected horizontal Peak Ground Acceleration (PGA) of 0.04g with a 10% exceedance in 50 years (the 475-year return period event). The Deterministic Seismic Hazard Analysis (DSHA) approach yielded a horizontal PGA of 0.04g, which is comparable to the PGA of the 475-year return period event of the PSHA. A scenario for the Maximum Probable Earthquake (MPE) could be represented by an event with M_W 7.4 at an epicentral distance of 238 km, causing a horizontal PGA of 0.04g and a vertical PGA of about 0.02g.

Earthquakes from the Hindu Kush subduction zone produce negligible effects at Jam in terms of seismic hazard due to the strong attenuation of ground motion because of the distance from Jam. This conclusion is valid for the PGA as well as for the spectral ordinates at the fundamental periods of vibration of the minaret (<1 sec).

The low seismic hazard at Jam is consistent with both the seismotectonic setting and the historical seismicity of the region. Jam lies in north-western Afghanistan in a zone characterised by a relative seismic quiescence. Though Jam is located in close proximity to the Herat (Hari-Rud) fault, a prominent strike-slip tectonic lineament in north Afghanistan, currently obtainable data appear to suggest that the geological feature is characterised by a very low level of activity, if any. This inference is complemented by historical seismicity dating back to 734 AD.

Finally, according to the Global Seismic Hazard Assessment Programme (GSHAP) in Continental Asia of Zhang et al. (1999), the 475-year return period PGA in the region of Jam is also estimated to be in the range of 0.04–0.08g.

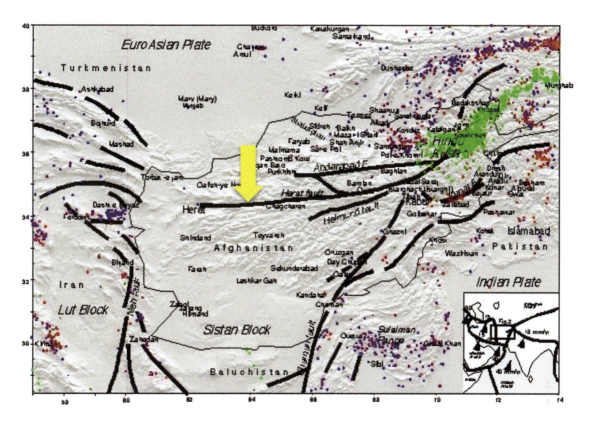

Figure 8. Historical and modern earthquakes in Afghanistan and major faults (from Ambraseys & Bilham, 2014). The yellow arrow indicates the position of Jam.

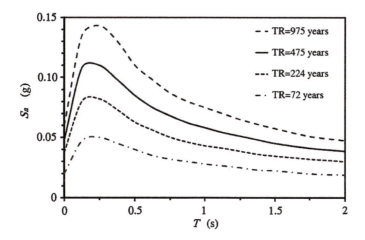

Figure 9. Elastic spectra at Jam site (from Menon *et al.*, 2004).

In Figure 9, the uniform hazard horizontal elastic acceleration spectra are shown for different values of the return period, TR. They were derived (Menon *et al.*, 2004) by means of the probabilistic approach. As one can see, for the return period of 475 years, the amplitude varies between 0.06g and 0.04g for periods between 1.0 and 2.0 seconds. Because of the historical importance of the structure, higher return periods should be used. A seismic analysis of the minaret was carried out by Menon *et al.* (2004) using the linear response spectrum analysis and taking into account the dynamic soil–structure interaction by inserting into the Finite Element (FE) model linear springs at the base.

Static and seismic analysis of the minaret have been performed by Clemente *et al.* (2014). They used and synthesised the stratigraphic and geophysical data collected by Orlando (2005) and Margottini (2005) before evaluating the seismic analysis and global stability analysis of the tower against soil collapse in the

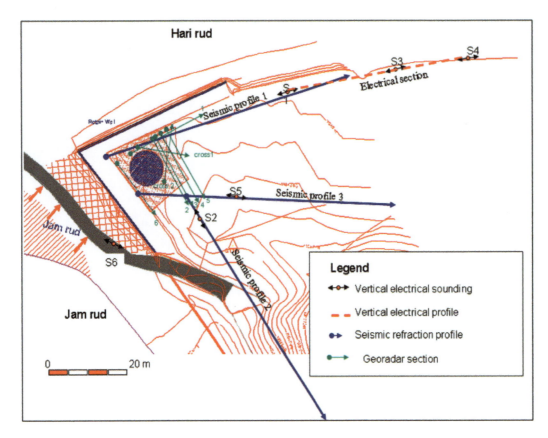

Figure 10. Geophysical surveys conducted in the vicinity of the Jam minaret.

present configuration, with the hypothesis of increasing bending moment at the base section, assuming elastic/perfectly plastic behaviour for the soil. Then, a finite element model was set up, which was used for a modal analysis and then for a seismic push-over analysis, based on both single and multi-modal approaches. Unfortunately, despite this very sophisticated analysis and elaboration, the limited availability of geotechnical data, especially for extreme conditions such as flood events and rising of the water table, did not allow firm conclusions to be drawn on this very important topic (Clemente et al., 2014).

5 GEOPHYSICAL SURVEYS

In March 2005, a geophysical campaign aimed at understanding the local stratigraphy and setting of the minaret foundation was performed. As usual, a multi-technique approach (Orlando, 2011) was employed, including georadar (Ground-Penetrating Radar, GPR), Seismic Refraction (SR) profiling and Vertical Electrical Sounding (VES). The GPR was applied with the aim of investigating the foundation soil surrounding the minaret, while SR and VES were applied with the aim of drawing a large-scale setting of the bedrock and the overlying deposits. The locations of the surveys are shown in Figure 10. The choice of surveys was a compromise between the objectives of the study and the difficulty in transporting heavy equipment from Italy to Jam.

5.1 *Ground-Penetrating Radar*

The aim of georadar was to investigate the possible extent and thickness of the minaret foundation, as well as the potential existence of buried archaeological remains. Details of the appropriate method can be found in Benson (1995) and Annan (2005). The GPR data were collected with a Sensors & Software instrument equipped with a 200 mHz antenna (Sensors & Software Inc., Mississauga, Canada). The data were processed with a time zero correction, a Butterworth bandpass filter and an exponential filter. The maximum depth

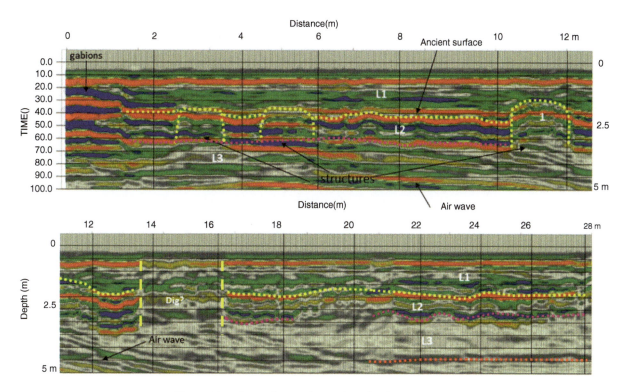

Figure 11. Georadar section 1 (location in Figure 10).

of investigation was 4–5 m, in which three layers have been clearly recognised (Figure 11). The first layer, starting from the topography, is about 1.5–2 m thick and it does not seem to include relevant anthropogenic structures. The second layer, with a thickness of 1–1.5 m, seems to include some anthropogenic structures and its top could be interpreted as an ancient topographical surface. The third layer, with its base at 4.5–5 m depth, is characterised by a strong attenuation of signal, probably due to an increase of water and/or clay content at that depth. This layer does not seem to contain structures. The analyses of the profiles acquired nearer the minaret do not show any evident anomaly which could be correlated to the foundation of the minaret inside the first 2–3 m from the topographic surface. In the past, the area near the minaret was probably attractive to excavations, as evidenced by the chaotic back scatters recorded in the georadar profile (Figure 11).

5.2 Seismic refraction

The seismic data were acquired with a Geode Seismograph instrument (Geometrics, San Jose, CA) equipped with 24 vertical geophones (7 Hz frequency). The geophones were spaced 3 m apart and seven shots were fired on each line. The inversion was based on the delay time (Gardner, 1939; Bernabini, 1965; Grant & West, 1965). The investigated area can be described with a three-layer model (Figure 12). Starting from the topographic surface, the first layer has a seismic P-wave velocity of about 0.5 km/s, the second a velocity of about 2.5 km/s, and the third a velocity of about 4.5 km/s.

The first layer, having a variable thickness from 2 to 10 m, can be correlated with the weathering layer, the second can be correlated with the sandy-gravel formation, and the third with the granite formation. Profile 3, collected near the minaret, detects the bedrock at a depth of about 20–25 m. The minimum depth of bedrock was detected by profile 2, acquired in the Jam Rud river along the valley. The seismic setting of the area is summarised by the section of profile 1, mapped in Figure 12. The zero of the section is very close to the minaret.

5.3 Vertical electrical soundings

Five vertical electrical soundings were collected, located at the same level as the minaret (Figure 10). The data inversion was based on least-squares optimisation methods (Lines & Treitel, 1984). The VES data show

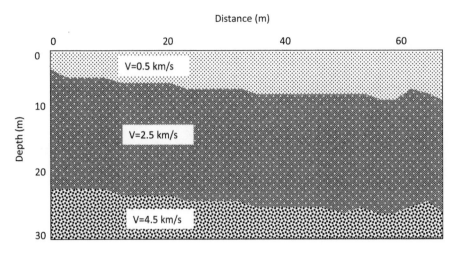

Figure 12. Seismic refraction of profile 1 (location in Figure 10).

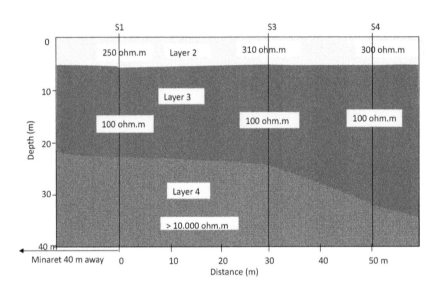

Figure 13. Electro-stratigraphy carried out from interpolation of VES S1, S3 and S4. The section is located parallel to the Hari river (location in Figure 9). S1 is about 40 m from the minaret.

that the investigated ground can be schematised by four layers: the first layer, starting from the surface, with a resistivity less than 80 ohm-metres ($\Omega \cdot$m); the second with a resistivity above 300 $\Omega \cdot$m; the third with a resistivity around 100 $\Omega \cdot$m; the fourth layer with a very high resistivity of more than 10,000 $\Omega \cdot$m.

Given the resistivity and the geology of the area, we can interpret the resistivity in terms of lithology as follows: the first layer can be formed by sandy-clay sediments, the second by sandy-gravel sediments located above the water table in the unsaturated zone, the third by sandy and gravel sediments located below the water table in the saturated zone, and the fourth layer by the granite formation. In Figure 13, the electro-stratigraphy of a profile obtained by interpolation of VES numbers 1, 3 and 4, located parallel to the river, is shown as an example. Layer 1 was not mapped in this section because of its small thickness (0.5–2 m).

The joint interpretation of vertical electrical soundings and seismic refraction data gives a good overlap in describing the configuration of the ground near the minaret. The first 6–8 m of the ground is described by VES with two layers and by SR with one layer. This portion of ground is unsaturated and formed by sediments with different silt contents. The near-surface layer probably consists of the latest alluvial deposits, as shown by GPR. Both seismic and SR profiling detect the layer between 6–8 m and 22–32 m depth. This layer is water-saturated and probably consists of sandy and gravel sediments. The bedrock that appears to be at about 20–25 m depth near the minaret is probably related to the granite formation outcropping in the area and is

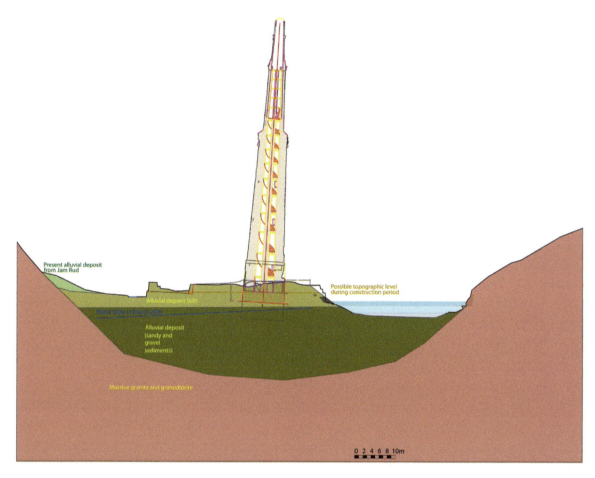

Figure 14. Geological section below the minaret, along the maximum direction of leaning.

detected by both VES and SR. A summary of all the available information allowing the realisation of the geological section is represented in Figure 14.

6 GEOLOGICAL EVOLUTION OF THE SITE IN HISTORICAL PERIOD AND RELATIONSHIP WITH MINARET INCLINATION

Not too much information is available on the history of the minaret and related damage. It is quite reasonable to say that the minaret was safely constructed, and only in a second stage, for many possible reasons, has it started to lean. In fact, in all high-rise historical monuments where there is a problem of verticality during the construction phase, the architects immediately rectify the upper part of them. This happened with the Pisa tower as well as other towers/minarets.

The first official record of the minaret (ICOMOS, 2002), dates from 1944, in the journal *Anis* of the Society for Afghan History. The minaret was rediscovered in 1957 by André Maricq (CNRS, France). A topographical survey was carried out in 1959 by J. Fischer and co-workers from the University of Cambridge, and in 1961–62 Andrea Bruno (Istituto Italiano per il Medio ed Estremo Oriente) conducted an architectural survey (measured drawings and proposal for restoration).

The first reinforcement of the base of the minaret took place in 1963–64, when a temporary stone and timber dam was built with the assistance of local villagers. Further surveys were carried out in 1971, 1973, 1974 and 1975 to determine the degree of leaning of the minaret. This was discovered to be moderate and not compromising the stability of the monument. Basic precautionary stabilisation measures were financed by UNESCO in 1978. These consisted of preventing the river from undermining the base of the monument by means of large stone-filled metal gabions. A number of priority tasks were identified, including regular inspection and maintenance of the gabions, sampling at the base of the minaret to determine the dimensions and structural characteristics of the foundation, an archaeological survey of an area of c. 40 m radius around the monument, monitoring of the amount of water flowing round the monument, consolidation and repair of the base, possible insertion of a reinforced concrete ring (dependent upon the results of the survey of the foundation), consolidation and counterbalancing of the structure by means of a balancing basin, and the provision of equipment and buildings. This work came to an abrupt end with the outbreak of the civil war.

In mid-1995, Dr A. W. Najimi visited the site and recommended that a gabion wall should be built to protect the minaret from flooding and erosion by the rivers (Najimi, 1995). It was not until 1999 that it was possible to build a wall of stone, reinforced with wood, along the Jam Rud, which only partially solved the problem of erosion. Later that year, following a further mission by Professor Bruno, a similar wall was built to mitigate undercutting of the base.

A group of experts that visited Jam in August 2001 recorded the fact that erosion was continuing, threatening the foundations of the monument. It was decided to remove the earlier wall, which was contributing to the accelerated erosion, and to create a new 45-metre-long wall of stone-filled gabions 10 m upstream along the bank of the Hari river; another, shorter wall was built alongside the Jam Rud. At the same time, dry stone masonry walls were built along other parts of the rivers to the same end.

Recently, after the civil war, technical missions were carried out by Professors Bruno and Borgia in March 2002, Professors Bruno and Margottini in 2004, and Professors Bruno, Margottini and Orlando in 2005. After these missions a severe flood occurred at the end of April 2007, destroying part of the gabions, close to the minaret. After this damage an immediate emergency intervention was planned by Professors Bruno, Fattorelli and Margottini and executed by local workers in October and November 2007 (Betastudio, 2007). It is not clear if this flooding had any consequence for the leaning of the monument.

Figure 15 represents the sequence described above, from 1961 until 2007. It is quite clear that the water was, until 2000, flowing very close to the minaret and producing an undercutting of the foundation and, potentially, a reduction of bearing capacity. Especially in 1961, the Jam Rud and the Hari Rud were just bordering the minaret.

It is clear from Figure 15 that the minaret has been recorded with severe leaning since, at least, 1974 (Bruno & Margottini, 2011). This leaning might be attributed to many causes but the most probable is related to river evolution and then the loss of bearing capacity due to flood events and fluvial erosion close to the foundation. Fluvial erosion is caused by rivers and streams, and can range from gradual bank erosion to catastrophic changes in river channel location and dimension during flood events. In order to prove such a hypothesis, a numerical elaboration of bearing capacity in various situations experienced by the monument was developed.

The ultimate bearing capacity of a shallow foundation can be defined as the maximum value of the load applied for which no point of the subsoil reaches rupture point (Frolich method), or else for which rupture extends to a considerable volume of soil (Prandtl method and subsequent). Many models and formulae have been developed since Prandtl's 1920 theory of plastic failure.

In the present chapter, an evaluation of ultimate bearing capacity for different situations is described. Many approaches have been used and reported in order to show the variability of different methodologies and the consistency of results.

Due to the unsecure condition of the site, where only in 2004 and 2005 was it possible to conduct appropriate field and geophysical investigations, the following assumptions were applied in the absence of any borehole and soil mechanics laboratory tests:

- the project load was considered only under drained conditions, even in the case of flooding, due to the high permeability of the alluvial deposit susceptible to dissipation of any sudden increase of pore pressure;
- the foundation was not identified with geophysical prospection, either because it is too deep for georadar or simply too close to the outer structure; foundation thickness was considered as running from just below the entrance level of the Jam minaret to the depth of the assumed water table at the time of construction (about the bottom of the river bed, at 6.5 m from the entrance level; see geological section); thus the thickness of the foundation was estimated at about 6.3 m;
- the foundation material was considered to be of high quality, composed of stone with good mortar, similar to the material identified in two different minarets in Herat (Margottini 2010; Urban, 2010);

Figure 15. The minaret and associated embankment in 1961, 1963, 1964–65, 1974–75, 1999 and 2005; in April 2007 after flooding destroyed the gabions; in November 2007 after the reconstruction of a protective wall (Bruno & Margottini, 2011).

– the alluvial material, where only a sedigraph analysis was performed, was characterised, in terms of shear strength, with no cohesion and a friction angle of 28°;
– the water table was considered in different states, according to rainy or flooding periods;
– the river bank was analysed at the time of construction, during erosional stages, and in present conditions after the construction of the gabions.

The geometric and mechanical properties of the monument were obtained from Santana Quintero and Stevens (2002), Stevens (2006), and Menon et al. (2004), as well as from direct investigations. They can be summarised as:

– Height from entrance level (about 4.5 m below topographic level): 65.0 m;
– Diameter at entrance level: 9.9 m;
– Strength of masonry based on 38 mm cube sample: 4.0 MPa (40,788 kg/cm^2; 4,000 kN/m^2);

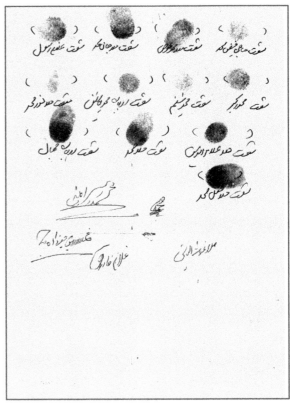

Figure 16. Permit of Prof. Andrea Bruno, authorising the inspection of the site of Jam and signed by both warring groups of Taliban and Mujahideen in 1999. The borderline of the war passed precisely through the Jam area (courtesy and © Andrea Bruno).

- Density of masonry: 1.6 t/m^3;
- Young's modulus of elasticity: 3,200 MPa;
- Inclination of the minaret: 3.4°;
- Maximum top displacement: 3.35 m;
- Total axial load: 32.630 kN (average gravity load 4.24 kg/cm^2; 415.80 kN/m^2);
- Present gravity load at footing level, after leaning, indicates 90 kN/m^2 in tensile border and 950 kN/m^2 in the compressive one;
- Bending moment at the base of the tower: 42.940 kNm;

Given the above assumptions and the geometric and mechanical properties of the minaret, a preliminary attempt to evaluate the ultimate bearing capacity (least gross pressure intensity which would cause shear failure of the supporting soil immediately below and adjacent to a foundation) in different assumed boundary conditions was implemented. Bearing in mind the difficulties of reaching the place in a region with attacks on UN cars and personnel (Figure 16), and then not being allowed to survey since 2005 (with all the limitations of a one-day helicopter mission), the lack of any proper stratigraphy (not a single drilling machine is available in Afghanistan for these materials) or laboratory data, and without a monitoring network, it is clear that some assumptions have to be made.

Cohesion was assumed as zero and friction angle as 28° for the grain-size distribution and classification of materials (Obrzud & Truty, 2012). A density of 17 kN/m^3 was calculated in the laboratory from sampled materials, but related only to the sand with silt deposits.

Different boundary conditions were considered, such as: the geometry and water table during the construction; a flood raising the water table to entrance level; with a flood depositing soil material until present topography and raising water table to the same; with the latter and embankment erosion; in the conditions

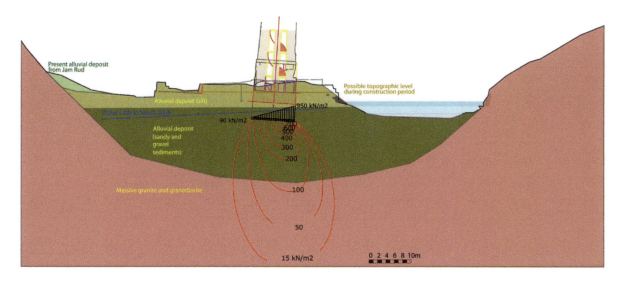

Figure 17. Non-uniform distribution of gravity load at the footing level and modification of the applied gravity load distribution at depth after the leaning of the minaret, according to Boussinesq theory.

recorded in 1974 with embankment erosion and the water table approximately 5 m below surface; as today with gabions.

The test showed that the worst conditions may appear for a combination of factors, which included the raising of the water table to surface level, typical during flood periods, especially when associated with embankment erosion.

Due to the 3.4° of leaning, the gravity load on the foundation plane of the minaret is now changing from an average stress of 415.80 kN/m^2 to a value close to zero in the tensile part at its footing (90 kN/m^2). The compressive strength at the other end of the cross section of the base is 950 kN/m^2 (Menon *et al.*, 2004). Figure 17 illustrates the stress due to the gravity applied load and its distribution at depth according to Boussinesq theory.

7 MONITORING THE INCLINATION IN THE LAST 15 YEARS

The Jam minaret and its immediate environment have been surveyed on a number of occasions over the past years. These surveys were designed to monitor the changes in the inclination, shape and position of the minaret. Under the local conditions in Jam, these surveys were extremely difficult to execute and only possible through the exceptional effort of conservation professionals.

The following assessment is based on three reports:

– October 2002 (Santana Quintero & Stevens, 2002);
– July/August 2006 (Stevens, 2006);
– November 2014 (Stevens, 2014).

7.1 *2002 survey*

This is the first modern survey after the end of the Afghan war. For the first time, two experts visited the site and conducted a survey in line with modern standards. According to Santana Quintero and Stevens (2002), the 2002 survey was carried out over a three-day period with a handheld Global Positioning System (GPS) receiver and a reflectorless Leica TCR 307 Total Station. In addition to this, a Nikon Coolpix 5500 digital camera with 5 MP resolution was used for a photogrammetric survey.

Two base points (ST101 and ST102) were established and surveyed with a handheld GPS receiver with an accuracy of not much better than 5 m. One of these points was reported to be 10 m and the other 300 m away from the minaret.

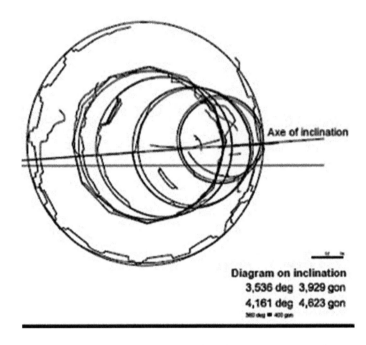

Figure 18. Central axis and angle of inclination of the minaret (Santana Quintero & Stevens, 2002).

The coordinate values of the two points were given as:

– East (M) 41 S 0639357, North (M) 41 S 3807168, Height 1905;
– East (M) 41 S 0639621, North (M) 41 S 3807244, Height 1888.

Because of the low accuracy of the equipment available for this survey, the coordinates of these two points cannot be used for deformation monitoring. Unfortunately, the photogrammetric survey referred to in the report is also of an accuracy below that required for the monitoring of deformations. This can be said because the photogrammetry was based on interior camera orientations which were derived from the nominal focal length of the camera.

Overall, the surveys were executed in a very limited time and the methods used were, with the exception of the Total Station survey, generally of sufficient accuracy for a topographical survey but, nevertheless, relatively low and not suitable for deformation monitoring. In 2014, only two of the 2002 survey points could still be used.

The method of producing sections through the minaret and determining its central axis and its angle of inclination (Figure 18) was very well designed and highly appropriate. What could be improved, marginally, is the way the axis and inclination were calculated.

7.2 *2006 survey report*

The 2006 survey (Stevens, 2006) makes reference to the 2002 survey; the same equipment is used and the report states that a similar inclination of the minaret was observed. The report also states that the difference between the two surveys was 5 mm. To obtain an accuracy of a few millimetres is not easy and the operational conditions, such as the steep sites, the lack of permanent targets on the horizontal sections and the uncertainty in the control network, suggest that the results are unlikely to be of millimetre-level accuracy. The report (Stevens, 2006) mentions primary and secondary reference points in relation to the 2002 survey; however, there is no mention of secondary points in the 2002 report (Santana Quintero & Stevens, 2002). There is also no reference to a precise GPS survey in the 2006 report. This would suggest, unfortunately, that any reference points were not positioned accurately as far as their absolute position was concerned, although the relative positions are determined with a considerably higher, though not quantified, accuracy with the Total Station instrument.

Table 1. Coordinates of two monitoring points installed on the minaret (Stevens, 2006).

UTM zone 41S	E0 63	N0 380	
	Easting (m)	Northing (m)	Height
Standpoint	9366.707	7137.946	1906.210
M1	9334.006	7187.699	1942.272
M2	9334.210	7189.344	1955.920

In 2002, two monitoring points were also installed on the minaret (M1 and M2) for future monitoring measurements (Santana Quintero & Stevens, 2002). Points M1 and M2 were positioned at heights of 40 and 54 m, respectively. The 2006 survey (Stevens, 2006) reports that the original point network was resurveyed, as some points had disappeared and some positions had physically changed. It would seem that points M1 and M2 were observed from one station only, referred to as "standpoint" in the report. This is not ideal for a monitoring survey but logistics in Jam are not easy. The coordinates of M1 and M2 and the "standpoint" were given according to Table 1.

7.3 October 2014 survey

The document of the October 2014 survey is not available but its conclusions are reported in Stevens (2014). According to the latter, a survey performed by an Afghan team in October 2014 indicated:

- a horizontal displacement at the top of the minaret of 3.35 m;
- an inclination angle of 3°25′27″.

According to the same Afghan team, the 2006 survey by Stevens indicates (data not found in Stevens, 2006):

- a horizontal displacement at the top of the minaret of 3.12 m;
- an inclination angle of 3°07′06″.

This led to this Afghan team's conclusion that the minaret's inclination had increased dramatically (by 24 cm) over the intervening eight years. Stevens (2006) stated that the horizontal displacement in 2006 was 3.128 m and the inclination 3.4405 decimal degrees or 3.8228 gradians (gons), which is the equivalent of 3°26′26″ and contrary to the figure reported above. The information spread that the minaret's inclination had increased significantly over the previous eight years, putting this World Heritage Monument in potential danger, was premature and apparently founded on an incorrect interpretation of the 2006 survey.

7.4 November 2014 survey

The November 2014 survey procedures are discussed in much more detail in Stevens (2014) by comparison with earlier exercises. However, there are still some uncertainties regarding the inclination of the central axis of the minaret and the conclusions reached.

The comparison of the Electronic Distance Measurement (EDM) measurements (with reflector and infra-red modes) of fixed points M1 and M2 in 2006 and 2014 (Figure 19) reveals a movement of, respectively, 16 and 28 mm. Given the heights of points M1 and M2 (40 and 54 m, respectively) the relationship between 16 mm and 28 mm is nonlinear; linearly, the figures should be, for example, 16 mm and 22 mm or 20 mm and 28 mm. This suggests a bending of the structure. However, it is possible that the survey accuracy is simply not good enough to draw conclusions of this nature. Figure 20 reports the results of November 2014's survey.

Figure 19. Fixed points M1 and M2 for monitoring by Total Station, used to compare 2006 and 2014 survey results (Stevens, 2014).

8 REINFORCEMENT OF BASEMENT

The outer layers of the masonry structure of Jam minaret have been partially lost over time, apparently due to forced removal and erosion. This could affect the overall stability of the entire structure and an emergency intervention was planned and executed in 2005 and 2006.

The surface concerned by the intervention is from approximately one meter below current surrounding ground surface, to approximately three meters above current ground surface. Original wall thickness is approximately two meters in the section concerned. This thickness is reduced by between 10 to 70 centimetres.

In order to reinforce the basement of the Jam minaret, the execution of masonry works at the base of the minaret was executed in 2005 and 2006. The new masonry, that follows the texture of the original one, has leaved the re-enablement of the building volume exactly like the original masonry surface. This new brick work has been keep in arrears of approximately 12 cm from the decorated wall surface, so as to leave also the possibility for a future completion that should be undertake on the frame of a full restoration program on the octagonal base and on the cylindrical trunk (Bruno, 2006).

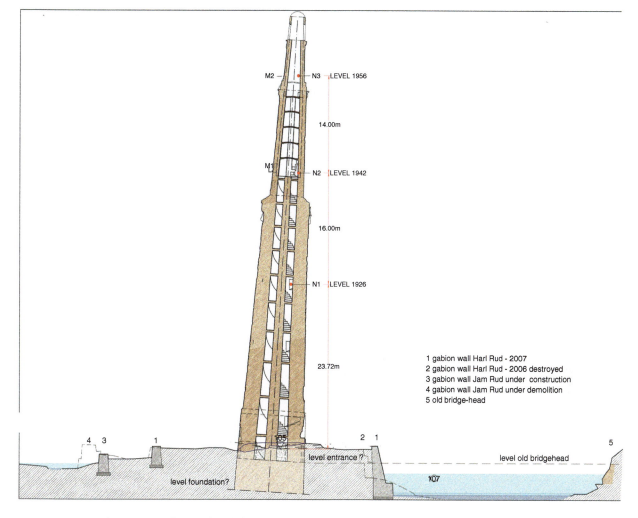

Figure 20. The results of the November 2014 survey (modified from Stevens, 2014).

The reconstruction of the masonry was realized using bricks produced on purpose in the brick-kiln in Herat, and placed in order to create a compact connection to the ancient masonry (Figure 21).

9 THE APRIL 2007 FLOOD AND EMERGENCY INTERVENTION

9.1 *Impact of the flood*

At the end of April 2007, one of the largest floods of the last 20 years occurred in the valley of Jam. According to eyewitnesses, the water affected the morphologically depressed part of the valley, persisting for a few days. Figure 22 shows a reconstruction of the flood.

One of the major problems affecting the stability of the minaret was the destruction of the gabions at the confluence of the Hari Rud and Jam Rud. This destruction could, theoretically, pose serious problems to the short-term conservation of the minaret, as described in previous sections. Figure 23 illustrates the destruction of these gabions by the flood.

An emergency plan was immediately established by the Afghan Minister of Information and Culture (MoC) and UNESCO, with the aim of immediate reconstruction of the gabions, before the next (2007/2008) winter season. The project was designed for UNESCO by Betastudio (2007) and provided to Afghan experts by the MoC in October 2007. Due to the difficulties of reaching the area, the project was implemented solely by the Afghan MoC, and only a detailed presentation of the construction need in Kabul offices was undertaken by UNESCO.

Figure 21. Reinforcement of Jam minaret basement in 2005 and 2006 compared with initial situation.

Figure 22. Extent of the end-of-April 2007 flood.

9.2 *Hydraulic analysis*

The carrying capacity of the flows along the course of the Hari Rud cannot be evaluated by taking an approach based on correlating those flows with meteorological impacts, either because the particular river system is seasonal rather than torrential in nature, or owing to the dearth of data available to permit a sufficiently reliable analysis. Similarly, it is not even possible to take a direct approach based on statistical analysis of its carrying capacity because the necessary observations are not available.

Thus, for the purposes of our hydraulic analysis, the carrying capacity was quantified from our knowledge, albeit approximate, of the water levels reached by the river at the end of the spring period, gathered from

Figure 23. Destruction of old gabions following end-of-April 2007 flood.

simple direct observations of the height difference between the river surface and the top of the river bank. Such observations, together with a survey carried out in the area surrounding the minaret, have been used to deduce the geometry of the river channel, its depth and the surface area of the water. The gradient can be calculated, with a higher degree of approximation, as equal to 0.01 (1%).

This information was used to calculate the hydraulic parameters that characterise the river in flood, on the assumption that the river has, approximately, a uniform regime.

The available data and photographs allowed grain-size distribution of the solid material that constitutes the gravel shore, which can be classified as a fine gravelly material, indicative of a relatively rough riverbed.

The geometry of the fluvial section has been deduced from the available maps and surveys (Santana Quintero & Stevens, 2002), and the results of a study on the consolidation of the minaret currently being carried out for UNESCO. This allows the level and height of the areas above river level (map of the flatland, distance between the river banks) to be inferred with precision, whereas some approximation was necessary as far as the submerged part of the main riverbed was concerned. The section near the minaret, just upstream from the confluence with the Jam Rud, was taken as a reference section for the hydraulic evaluations. These evaluations were used to construct two significant hydrometric scenarios, Cases 1 and 2.

Case 1 concerns a hydrometric scenario such as that observed in the flood at the end of April 2007, when the river was roughly 3.0–3.5 m below the top of its banks, while Case 2 concerns a hypothetical situation, the nearly maximum level that could be contained by the banks in particularly extreme hydrometric conditions. In Case 1, involving the hydrometric conditions observed at the end of April 2007, the evaluations gave an estimated discharge of 400 m^3/s. In Case 2, the discharge estimate was 700 m^3/s. Table 2 shows the hydraulic parameters that characterise the flow in both situations. Attention is drawn to the average velocity, which varies from 3.0 to 4.0 m/s, to the shear, which results in the dense solid matter carried that can be seen during the spring spate, and to the erosion of the banks, which have all sapped the existing river defences.

Conclusions about the recommended project, materials and implementation procedures were then provided to the Afghan MoC by UNESCO consultants (Betastudio, 2007). Figure 24 illustrates the solution selected for the reconstruction of the entire embankment.

After only two months, in December 2007, the entire embankment had been fully reconstructed, and was in a safe condition. The only difficulties encountered were in digging out the big block of rock on the river bed, necessary to the realisation of the diaphragm shown in Figure 23. This element was postponed, and the

Table 2. Hari Rud: hydraulic parameters in flooding conditions.

Hydraulic parameter	Case 1	Case 2
Q Total (carrying capacity) [m^3/s]	400.00	700.00
Min Chl Elev (minimum channel elevation) [m a.s.l.]	1,896.50	1,896.50
W.S. Elev (water surface elevation) [m a.s.l.]	1,900.11	1,902.09
Max Chl Dpth (maximum channel depth) [m]	3.61	5.59
E.G. Slope (energy gradient) [m/m]	0.010	0.010
Top Width (width of flow section) [m]	35.35	44.09
Flow Area (flow section) [m^2]	116.79	174.18
Chl Dpth (channel depth) [m]	2.70	3.95
Avg. Vel. (average velocity) [m/s]	3.14	4.02
Vel. Head (kinetic energy) [m]	0.50	0.82
Shear (tangential force) [N/m^2]	254.13	367.22

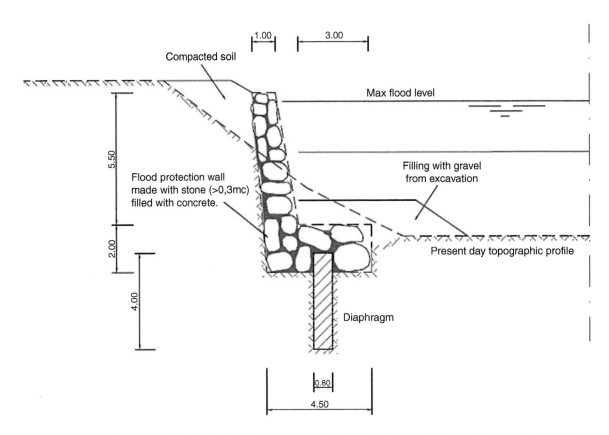

Figure 24. The proposed solution for the reconstruction of the entire embankment (Betastudio, 2007)

final diaphragm, just in front of the present gabion wall, will be implemented with more advanced equipment in the near future. Figure 25 shows the excellent final result implemented by the Afghan MoC.

10 FURTHER NEEDS

The investigations and the emergency intervention executed up until now are not sufficient to evaluate and ensure the medium- to long-term stability of the Jam minaret. As a consequence, an investigation plan for future interventions, together with immediate short-term emergency conservation, was drawn up in 2015

Figure 25. The Jam protection gabions before the flood destruction in April 2007 (left) and after the reconstruction in October/November 2007 (right).

(Bruno *et al.*, 2016). Such activities should be implemented as soon as the safety conditions and logistics will allow a permanent presence at the site. These include:

- A Total Station survey of network and horizontal sections, to monitor the minaret and to quantify absolute horizontal and vertical displacements of the entire structure, as well as deformations and changes in inclination of the minaret, with an accuracy of a few millimetres.
- Laser scanning, to acquire a point cloud covering the entire surface of the minaret and, potentially, its inside too. This will make it possible to analyse the entire surface, discover dislodged stones through a repeat scan and monitor deformations of the surface with an accuracy of 5 to 15 mm depending on surface condition and position in the minaret (higher points are less accurate).
- Digital photogrammetry, for texture mapping of the minaret surface and for photogrammetric modelling of small subsections of the minaret.
- Electric tomography, to reconstruct the 3D distribution of the bedrock as a support to proper design.
- Drilling and undisturbed soil sampling (depending on available equipment), to improve understanding of the profile of the subsoil and quantify geomechanical parameters.
- Reconstruction of the geotechnical conditions of the site, to define a geologic model relevant to a proposed project that includes pertinent engineering aspects. The geology must be expressed in quantitative terms that explicitly describe complexity and uncertainty to have value in engineering projects, particularly those that employ Reliability-Based Design (RBD) (Burland, 1987; Keaton, 2013).
- Synthetic Aperture Radar Interferometry (InSAR) with the Persistent Scatterer Pairs (PSP) approach; repeat-pass satellite InSAR is a very effective technology for measurement of ground deformations due to subsidence, landslides, earthquakes and volcanic phenomena. This technology allows terrain displacements to be measured with millimetric accuracy by processing data acquired from satellites orbiting more than 600 km above the Earth (Alberti *et al.*, 2017). Satellite data can also be used for the detection of illegal excavation.

Together with the above investigations, some short-term conservation activities can be implemented as soon as the area is safe again. These include:

- Hydrological and hydraulic interventions; by comparing the available photographs, it has been noticed that an intervention on the Jam Rud, tributary of the Hari Rud, was implemented after the realisation of the emergency intervention in 2007. Such stone wall filled with concrete it seems to have been carried out in several stages. Part of this wall later collapsed for an estimated length of about 20 to 30 m. After this collapse, a second gabion wall parallel to the previous one was erected, with a length of about 100–150 m (Source: Afghan Ministry of Information and Culture). It is very likely that the construction of this wall, which has greatly reduced the hydraulic section of the Jam Rud, has been the trigger for the

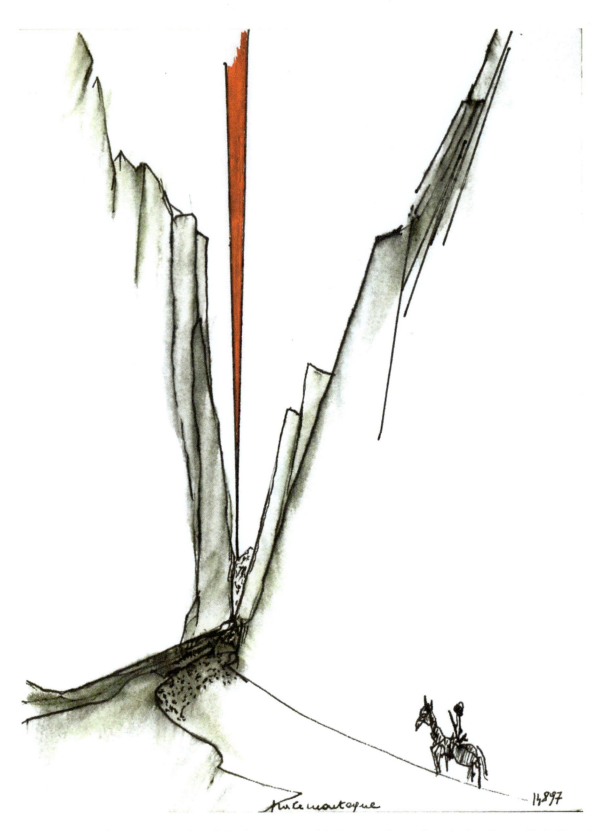

Figure 26. An artistic reconstruction of the site entrance with the Jam minaret from Andrea Bruno (© A. Bruno)

subsequent damage. The hydraulic analysis previously conducted by Betastudio (2007), by implementing a one-dimensional hydrometric HEC-RAS software model, hypothesised a flow rate of 300 m^3/s or even more, with a topographic level around 1,900 m above sea level (a.s.l.). A better discharge will be required after a proper hydraulic investigation. A first hypothesis can consider the widening of the left hydraulic (in the case that the wall gabions remain as they are now) for a width sufficient to always ensure the stability of the bank. The Jam Rud also has a very high fall, of the order of a few percent; it may be appropriate to realise, upstream, some gabion dams, with the aim of reducing the fall and slowing the water velocity and the transport of solids.

– Minaret decoration conservation; such activities are most essential because the damage can become irreversible. The field work should be developed in close cooperation with local authorities and experts.

11 SUMMARY

The Minaret of Jam has been a focus of conservation and research for over 45 years (Bruno, 2003b) and, in 2002, UNESCO declared Jam Afghanistan's first World Heritage Site.

Jam is located in the Ghur province of central Afghanistan, an inaccessible mountainous region about 260 km east of Herat, and the minaret is merely the most visible element of the surrounding archaeological site, largely uninvestigated.

The minaret is composed of four tapering cylindrical shafts. The tower structure, constructed with bricks and lime mortar, is composed of two parts: the external one, made of four tapered truncated blocks of ring-like cross section, one upon the other, and the internal one, composed of tapered blocks of circular cross section, which ends at a height of 43.50 m. Two spiral staircases are between the external and internal structures and also join them. The double helicoidal staircase runs from the base to the first cylindrical tier, leading to the top.

The local geology is strongly affected by the mountain environment, with a bedrock composed of massive granite/granodiorite, and covered by a few orders of alluvial deposit, characterised by sediments, related to different alluvial stages, and mainly composed of sand and gravel, heteroclastic and heterometric with an abundant presence of sand matrix.

A geophysical campaign, incorporating georadar, seismic refraction profiles and vertical electrical soundings (VES), provided information about the local stratigraphy and the typology of the foundation. It also highlighted the presence of buried archaeological remains at about 2.0 m from the ground surface. The seismic refraction analysis indicated that the soil could be schematised in three layers. The first layer, starting from the topographic surface, has a variable thickness of 2–10 m and can be correlated with the modern alluvial deposit; it exhibits a seismic wave velocity of about 0.5 km/s and its components in the field have been identified as silt, sand and gravel. The second layer can be correlated with a gravel and sandy formation, also from alluvial deposition, and shows a seismic wave velocity of about 2.5 km/s; large boulders belonging to such an alluvial deposit have been found during embankment reconstruction in 2007. The third layer, associated with the granite formation, exhibits a velocity of about 4.5 km/s. The bedrock near the minaret is at a depth of about 20–25 m. These results were confirmed by the VES.

The existing geotechnical data needs significant improvement through proper investigation, but a drilling machine suitable for work in silty-clay alluvial deposits with a large percentage of boulders or in unconsolidated conglomerate is not available in Afghanistan.

The tower suffers an inclination of about 3.4° north-north-eastwards, still not completely explained. It is clear that the leaning has occurred post-construction because no architectural verticality adjustment, as occurred in the Pisa tower, has been implemented. Total Station monitoring data from 2002, 2007 and 2014 cannot confirm increased leaning in recent times.

According to available data and modelling, the leaning can, most probably, be ascribed to undercutting of the foundation by the Hari Rud river and, possibly, to the contemporary reduction of bearing capacity during severe floods and raising of the water table.

As a consequence, after the severe floods that affected the site in 2007 and destroyed the gabions below the minaret, an emergency intervention was implemented by UNESCO and, at the end of 2007, the protective wall close to the Hari Rud was reconstructed with modern and safer designs. The Jam Rud river must yet be secured in similar fashion.

Emergency intervention to protect the masonry and, principally, the external decorations was implemented by UNESCO in 2005–2007, with important results for the long-term conservation of the monument.

All the activities described in this chapter have been the result of a continuous balance between the need for emergency conservation work and the availability of data and its interpretation. The latter are dependent on access and logistics to the site, in a conflict area where, in addition, any technological and geotechnical equipment must be imported from abroad.

A plan for future interventions was proposed in 2015 (Bruno *et al.*, 2016) and included further geological, geophysical and geotechnical investigations to allow proper hydraulic protection and stabilisation of the minaret foundation. These investigations and future works must be accompanied by an appropriate monitoring plan. Finally, on the basis of available information, knowledge of the site and the professional expertise of the authors (Bruno *et al.*, 2016), some short-term emergency interventions have been proposed and include the hydraulic securing of the Jam Rud river and completion of the stabilisation of the marvellous external decorations. Clearly, any possibility of intervention on site will depend on accessibility to the minaret, which is strictly dictated by the security status of the area.

REFERENCES

Abdullah, S. (1981). Geological observations & geophysical investigations carried out in Afghanistan over the period 1972–1979. In: Gupta, H.K. & Delany, F.M. (Eds.), *Zagros – Hindu Kush – Himalaya – Geodynamic evolution* (pp. 75–86). Washington, DC: American Geophysical Union.

Abdullah, S. (1993). Seismic hazard assessment in the Islamic state of Afghanistan. In: McGuire, R.K. (Ed.), *The practice of earthquake hazard assessment* (pp. 284). IASPEI/ESC Publications.

Alberti, S., Ferretti, A., Leoni, G., Margottini, C. & Spizzichino, D. (2017). Surface deformation data in the archaeological site of Petra from medium-resolution satellite radar images and SqueeSAR™ algorithm. *Journal of Cultural Heritage*, 25, 10–20.

Ambraseys, N.N. & Bilham, R. (2014). The tectonic setting of Bamiyan and seismicity in and near Afghanistan for the past 12 centuries. In: Margottini, C. (Ed.), *After the destruction of giant Buddha statues in Bamiyan (Afghanistan) in 2001* (pp. 101–152). Berlin, Germany: Springer-Verlag.

Annan, A.P. (2005). Ground penetrating radar. In: Butler, D.K. (Ed.), *Near-Surface Geophysics* (pp. 357-438). Tulsa, OK: Society of Exploration Geophysicists.

Bernabini, M. (1965). Alcune considerazioni sui rilievi sismici a piccole profondità. *Bollettino di Geofisica Teorica e Applicata*, 26, 106–118.

Benson, A.K. (1995). Applications of ground penetrating radar in assessing some geological hazards: Examples of groundwater contamination, fault, cavities. *Journal of Applied Geophysics*, 33, 177–193.

Betastudio. (2007). *Safeguarding of the Minaret of Jam in Afghanistan*. UNESCO project.

Borgia, A. (2002). *Preliminary geological hazard and foundations assessment at the Minaret of Jam, Afghanistan*. UNESCO internal report.

Bruno, A. (1963). Notes on the discovery of Hebrew inscriptions in the vicinity of the Minaret of Jam. *East and West*, 14, 206–208.

Bruno, A. (2002). *Afghanistan. Minaret of Jam. Mission March 2002*. UNESCO internal report.

Bruno, A. (2003a). *Compte rendu des missions (1999, 2003). Consideration sur la missione aout 2003 à Jam*. UNESCO internal report.

Bruno, A. (2003b). "The Minaret of Jam, Afghanistan", *World Heritage*, 29, UNESCO Publishing and Ediciones San Marcos.

Bruno, A. (2004). *UNESCO mission. February 21st–March 3rd 2004 at Jam and Heart*. UNESCO internal report.

Bruno, A. (2005). *Considerations sur la mission octobre 2005 à Jam. Minaret de Jam – Afghanistan*. UNESCO internal report.

Bruno, A. (2006). *Emergency and Restoration of Monuments in Herat and Jam: mission report. UNESCO internal report.*

Bruno, A. & Margottini, C. (2011). The Jam minaret in Afghanistan: Emergency works and long term stabilisation. In: Iwasaki, Y. (Ed.), *Proceedings of the special session on Geo-Engineering for Conservation of Cultural Heritage and Historical Sites. 14th Asian Regional Conference, ISSMGE 26 May 2011, Hong Kong, China.*

Bruno, A., Margottini, C. & Ruther, H. (2016). *Jam action plan 2015*. UNESCO internal report.

Burland, J.B. (1987). Nash Lecture: The teaching of soil mechanics – A personal view. In: *Groundwater effects in geotechnical engineering, Vol. 3. Proceedings of 9th European Conference on Soil Mechanics and Foundation Engineering* (pp. 1427–1441). Rotterdam, The Netherlands: Balkema.

Clemente, P., Saitta, F., Buffarini, G. & Platania, L. (2014). Stability and seismic analyses of leaning towers: The case of the minaret in Jam. *Structural Design of Tall and Special Buildings*, 24(1), 40–58. doi: 10.1002/tal.1153

Gardner, L. (1939). An area plan of mapping subsurface structure by refraction shooting. *Geophysics*, 4, 247–250.

Grant, F. S. & West, G. F. (1965). *Interpretation theory in applied geophysics*. New York, NY: McGraw-Hill.

Keaton, J.R. (2013). Engineering geology: Fundamental input or random variable? In: Withian, J.L., Phoon, K.K. & Hussein, M.H. (Eds.), *Foundation engineering in the face of uncertainty* (pp. 232–253). Reston, VA: American Society of Civil Engineers.

ICOMOS (2002). *Minaret of Jam (Afghanistan)*. Advisory Body Evaluation, rev. 211, (http://whc.unesco.org/archive/advisory_body_evaluation/211rev.pdf.)

Lines, L.R. & Treitel, S. (1984). Tutorial: A review of least-squares inversion and its application to geophysical problems. *Geophysical Prospecting*, 32, 159–186.

Margottini, C. (2005). *Engineering geology aspects related to the consolidation of the consolidation of the Jam minaret (North-Western Afghanistan). Final report on the UNESCO mission on Jam (North-Western Afghanistan) on 23 May – 2 June 2005*. UNESCO internal report.

Margottini, C. (2010). *Emergency Consolidation of the 5th Minaret in Herat (Afghanistan)*. UNESCO internal report.

Maricq, A. & Wiet, G. (1959). *Le Minaret de Djam: La découverte de la capitale des sultans Ghorides (XIIe–XIIIe siècles)*. Paris, France: Délégation archéologique française en Afghanistan.

Menon, A. & Lai, C.G. (2004). Seismic hazard assessment of Jam in Afghanistan. *13th World Conference on Earthquake Engineering, Vancouver, B.C., Canada, 1–6 August 2004*. Paper No. 2314.

Menon, A., Lai, C.G. & Macchi, G. (2004). Seismic hazard assessment of the historical site of Jam in Afghanistan and stability analysis of the minaret. *Journal of Earthquake Engineering*, 8(Special Issue 1), 251–294.

Najimi, W.N. (1995). *Jam, An Assessment Mission Report*, UNESCO internal report.

Obrzud, R. & Truty, A. (2012). *The hardening soil model – A practical guidebook*, Z Soil.PC 100701 report, revised 31.01.2012.

Orlando, L. (2005). *Geophysical survey in the Minaret of Jam, Afghanistan*. UNESCO internal report.

Orlando, L. (2011). Multidisciplinary approach to a recovery plan of historical buildings. *International Journal of Geophysics*, 2011. doi:10.1155/2011/258043

Santana Quintero, M. T. & Stevens, T. (2002). *Report: Metric survey tools in recording: Mussalah Complex Herat and Minaret Jam, Afghanistan*. UNESCO internal report.

Stevens, T. (2006). *Monitoring the leaning of Jam minaret*. UNESCO internal report.

Stevens, T. (2014). *Jam minaret. Assessment of inclination and immediate threats*. UNESCO internal Report.

Urban, T. (2010). *Archeological excavation in Herat*. UNESCO internal report.

Zhang, P., Yang, Z.-X., Gupta, H.K., Bhatia, S.C. & Shedlock, K.M. (1999). Global Seismic Hazard Assessment Program (GSHAP) in continental Asia. *Annali di Geofisica*, 42(6), 1167–1190.

Geotechnics and Heritage: Historic Towers – Lancellotta, Flora & Viggiani
© 2018 Taylor & Francis Group, London, ISBN 978-1-138-03272-9

Author index

Brandl, H. 145
Bruno, A. 229
Burland, J.B. 59

Ceroni, F. 99

de Silva, F. 99

Flora, A. 5
Fukuda, M. 191

Gottardi, G. 73

Harris, D.I. 59

Iwasaki, Y. 191

Lancellotta, R. 5, 39
Leoni, M. 15
Lionello, A. 73
Lisyuk, M. 171

Mair, R.J. 59
Marchi, M. 73
Margottini, C. 229
Montanari, T. 1

Orlando, L. 229

Puzrin, A.M. 123

Rossi, C. 73

Sabia, D. 5, 39
Shashkin, A. 171
Shashkin, C. 171
Sica, S. 99
Silvestri, F. 99
Spizzichino, D. 229
Squeglia, N. 15
Standing, J.R. 59

Ulitsky, V. 171

Viggiani, C. 5, 15

GEOTECHNICS AND HERITAGE: HISTORIC TOWERS